BOSTON STUDIES IN THE PHILOSOPHY OF SCIENCE
VOLUME XVIII
PHILOSOPHICAL PROBLEMS OF MODERN PHYSICS

BOSTON STUDIES IN THE PHILOSOPHY OF SCIENCE

VOLUME XVIII

PHILOSOPHICAL PROBLEMS OF MODERN PHYSICS

SYNTHESE LIBRARY

MONOGRAPHS ON EPISTEMOLOGY,

LOGIC, METHODOLOGY, PHILOSOPHY OF SCIENCE,

SOCIOLOGY OF SCIENCE AND OF KNOWLEDGE,

AND ON THE MATHEMATICAL METHODS OF

SOCIAL AND BEHAVIORAL SCIENCES

Managing Editor:

JAAKKO HINTIKKA, *Academy of Finland and Stanford University*

Editors:

ROBERT S. COHEN, *Boston University*

DONALD DAVIDSON, *Rockefeller University and Princeton University*

GABRIËL NUCHELMANS, *University of Leyden*

WESLEY C. SALMON, *University of Arizona*

VOLUME 95

BOSTON STUDIES IN THE PHILOSOPHY OF SCIENCE

EDITED BY ROBERT S. COHEN AND MARX W. WARTOFSKY

VOLUME XVIII

PETER MITTELSTAEDT

PHILOSOPHICAL PROBLEMS OF MODERN PHYSICS

D. REIDEL PUBLISHING COMPANY

DORDRECHT - HOLLAND/BOSTON - U.S.A.

Cloth edition: ISBN 90 277 0285 3
Paperback edition: ISBN 90 277 0506 2

PHILOSOPHISCHE PROBLEME DER MODERNEN PHYSIK
First published in 1963 by Bibliographisches Institut, Mannheim
Translated from the revised fourth edition by Dr. W. Riemer
and revised by R.S.C.

Published by D. Reidel Publishing Company,
P.O. Box 17, Dordrecht, Holland

Sold and distributed in the U.S.A., Canada and Mexico
by D. Reidel Publishing Company, Inc.,
Lincoln Building, 160 Old Derby Street, Hingham,
Mass. 02043, U.S.A.

Printed in The Netherlands

Professor Peter Mittelstaedt is a physicist whose primary concern is the foundations of current physical theories. This concern has made him, through his prolonged, incisive and detailed examinations of the structures and overall characteristics of these theories, into a philosopher of physics — of contemporary physics, to be precise, of relativistic theories of space and time, and of the logic of quantum mechanics, in particular. The present book, which expounds his main ideas in these matters, has seen four editions (in German), each including newer results — as indeed does the present translation: see the author's 1975 preface to the English translation.

Perhaps this is the place to repeat the author's chief problem and mention his own approach, even though they are expounded in his Introduction. How close is Mittelstaedt to Kant's understanding of science? We are at liberty to choose a framework for thought — a logic and a methodology — prior to experience (in the classic sense, to think *a priori*); yet we choose a framework so as to fit our empirical findings. How is this done? How may it be understood and justified? This is obviously *the* question of all philosophies that evolve from, and are in reaction to, Kant's system. Reminiscent of the philosophical method of Meyerson, Mittelstaedt stresses, first and foremost, that a discussion of this question of the *a priori* character of theories cannot be fruitfully conducted *a priori*; it must utilize detailed and precise knowledge of both available frameworks and available scientific contents, from scientific theories to empirical facts of nature, and with mastery of modern logic. It is in relation to these facts, or more precisely, to the specific problem of the process of measurement, that the author builds his theories. He contends that physical theories situate themselves conceptually; that they include their own ideas on the process of measurements; and that these (whether explicit or newly-made-so) offer the clues as to the frameworks best fitting them. Needless to say, this is a Kantian philosophy, yet inconceivable to Kant or to any thinker prior to the students of relativity and quantum theory.

Mittelstaedt's theory of scientific concepts and principles is not restricted

to his valuable and original work on measurement. This monograph has been
widely received in the German-speaking world and elsewhere, as a clear-
headed exposition of the wider problems and interpretive arguments of
contemporary epistemology of physics. We expect that his book will stimu-
late further study and critical debate in its revised and expanded incarnation
in English.

ROBERT S. COHEN
MARX W. WARTOFSKY

Center for Philosophy and History of Science
Boston University
August 1975

TABLE OF CONTENTS

PREFACE TO THE ENGLISH EDITION

By modern physics one usually means the period of physics that started with Einstein's analysis of space and time (1905) and that reaches to the present. Apart from the great accumulation of factual knowledge, two theories essentially distinguish modern physics and its entire structure from all preceding epochs of physics: the theory of relativity and quantum theory. The philosophical problems that have arisen in connection with these two theories constitute the subject matter of this book.

In the formulation of the problems, I took special care to do justice to the often conflicting approaches of physics and philosophy by a careful analysis of the conceptual bases of these two points of view. In so doing it was necessary to clarify in what way modern physics, an empirical science, can lead to philosophical questions and give rise to propositions on time, causality and logic. On the other hand, it was necessary to note that the possibility of obtaining truths not derived from experience, repeatedly presented since the beginning of Greek philosophy, still exists and has not been compromised in any way by the development of modern empirical science.

The book has evolved from lectures and seminars given at the Universities of Munich and Cologne during the years 1962–1974. Since the first edition, the text has been revised and enlarged in many respects. The present English edition differs from the fourth (German) edition, apart from minor improvements in the Chapters I, IV and V, mainly in the incorporation of some new results in Chapter VI, which made it possible to derive a complete calculus for quantum logic in the framework of the operational foundation of logic — without any recourse to the lattice of subspaces of Hilbert space. In order to give an adequate presentation of this result, a revision of major parts of Chapter VI was required.

During the preparation of the English edition of this book, especially in reformulating Chapter VI, I have been helpfully assisted by Drs. E. Drope and E. W. Stachow. The permanent interest in the realisation of the English translation and the careful editorial revision of the entire manuscript by the editor of the *Boston Studies in the Philosophy of Science,* Prof. R. S. Cohen is gratefully acknowledged.

PETER MITTELSTAEDT

October 1975

INTRODUCTION

Physics is a science that has a fixed methodological framework. Regardless of the specific content of an investigation, the way in which problems are posed and the methods by which they are solved are firmly established. The methods can be learned and, in principle, can be applied to any unresolved question. The results obtained in this way are always answers that are given in a clearly defined manner to questions formulated and posed by physics with similar clarity. In some respects the range of problems that can be confronted by this science is therefore already fixed by its methodological framework.

This characteristic appears to immediately exclude the possibility of philosophical problems within physics. For if by philosophy we understand a way of thinking that is open with respect to methodological and intentional constraints, then philosophical questions are not possible within the bounds of science. The constant challenging of these boundaries, which distinguishes philosophy from every science, cannot itself be a component part of one of these sciences.

If by physics we understand a discipline that examines phenomena under the constraint of established aspects, then philosophical reflection within physics can have only these very aspects as subject matter. Progress in physics will therefore never lead to new and previously unknown philosophical questions. Only if the intention of physics changed would certain philosophical judgements about physics become unsound. But as long as such change in the concept of physics is carried out on the basis of autonomous philosophical considerations, there is no reason to speak of philosophical problems of physics. The problems would rather have their source in considerations that are outside of physics.

It would be a completely different situation if the changes in the methodological framework of physics were made on the basis of physical, rather than philosophical, results. One could then justifiably speak of philosophical problems of physics. The physical results would then lead to questions that could no longer be resolved within the framework of physics.

The actual situation of modern physics corresponds precisely to this. Since the beginning of this century the methodological basis of physics has twice been changed fundamentally by experimental results: The first of these changes led to the theory of relativity, the second to quantum theory. The conceptual frame of physics was then again stabilized in a new, modified form by the establishment of these theories. Those philosophical issues which deal with the transformation of classical physics into the modern theories of relativity and quantum physics, will constitute the subject matter of this book.

The involvement with philosophical problems of physics was concretely occasioned by certain assertions first made in physics at the beginning of this century. The issue was that certain structures, whose validity no one had questioned until then, were declared to be false. Euclidean geometry, the law of causality, and logic were the most striking examples. From the physical point of view the justification for these assertions were to be found in experience, the only legitimate source of physical knowledge.

Against these assertions, but especially against their empirical basis, some serious philosophical objections can be raised. Every fixed methodological framework, by which a certain science is defined, uses certain categories, concepts and schemata for ordering and interpreting phenomena. Certain aspects by which the perceived can be considered are excluded in this way, yet other features are given prominence. Already from these schemata of ordering, by which perceptions are interpreted, it follows, in general without empirical cognition, that certain properties can be attributed *a priori* to all objects of the science in question. The most precise presentation of these ideas for science is found in Kant, who challenges the assertion of Hume that all knowledge must derive from experience. Kant includes the following in the knowledge that is *a priori* valid in this way for all objects of science: the structure of space and time, Euclidean geometry, the law of causality and the law of conservation of matter. A confrontation with the physical assertions mentioned above, on the invalidity of these structures will therefore necessitate a discussion of Kant's arguments for the possibility of *a priori* knowledge. The arguments given in modern logic for the *a priori* validity of the laws of logic have a similar basis. The conditions under which propositions about what exists (*Seiendes*) can at all be made, already determine the rules which these propositions must necessarily obey.

Since the *a priori* validity of certain structures of experience can be thus clarified, preference to experimental results becomes meaningless whenever it is precisely these structures that are to be examined. Physics complies with this situation inasmuch as it incorporates the *a priori* structures derived from the conditions of the possibility of experience, so that concrete laws of nature do not contradict these structures. This is not always apparent: All propositions of physics derive from experience and relate to situations that can be realized experimentally. But that does not imply that physics is based solely on experimental data. The *a priori* structures, it is true, are usually not emphasized as such, but rather appear in conjunction with contingent laws of nature and are therefore often almost hidden. Thus, for example, the law of causality is never formulated in physics as a separate law. The reason for this is that the causal law is already contained in the concrete empirical laws. Thus only certain types of differential equations have 'causal solutions', and it is only these equations that are used in physics.

In a similar way other *a priori* knowledge is also accommodated in physics. Certain general structures of the laws of nature ensure that in concrete cases the relevant *a priori* assertions are always fulfilled. Since the *a priori* structures which thus result from critical reflection on the conditions of the possibility of experience are accommodated in the development of physics, they cannot easily be refuted by physical experiments. The question therefore arises, what, under these circumstances, is meant by the assertion that the *a priori* structures may be doubted on empirical grounds?

On the basis of these considerations a change of those propositions that must be valid *a priori* for all experience is possible only if the conditions of the possibility of cognition, and thus the methodological framework and with that the very concept of physics, are changed. Such a change in the conceptual basis of experience would then also modify the structure of things that appear in experience. The recognition that changes of the *a priori* structures of physics are based on experience, must therefore not be understood in an empiricist sense. These changes derive rather from a change in the methodological framework of physics, which in turn occurred not because of philosophical considerations but exclusively for empirical reasons. The question, how empirical results can at all influence the choice of the categories with the aid of which these experiences come about, is the specific problem that modern physics poses for a philosophical interpretation of its results.

This question can only be answered by an exact analysis of the conditions under which empirical-physical results are obtained: the laws that govern the measuring processes. The fact, that all measuring instruments are also objects of experience and as such must obey the partly empirical laws of physics, has the consequence that these laws in turn can influence the structure of the very experience which can be obtained with the aid of measuring devices. The exact form of the laws of nature, which are at the same time the laws of the measuring device, then leads to the changes of the categorical framework, characteristic of modern physics, and thereby to the concept of experience relevant for this physics.

Under these circumstances there are fundamentally two methods of formulating physics as a theory of events of nature. One method starts with the requirement that the theory can only contain quantities that can be observed with actually existing measuring devices. Physics is then a theory of observable quantities, which describes nature as it reveals itself to real measuring devices. Such a theory exclusively uses categories for the ordering of phenomena that can be derived from a theory of the real measuring process. In addition there is the possibility of retaining schemata and categories for the ordering of phenomena that are familiar from classical physics, Euclidean geometry and classical logic without questioning the empirical realizability of these concepts. A physics constructed on this basis must, however, also contain a correct description of empirical phenomena, including the laws of the measuring devices. By the application of this knowledge one can then show, however, that the original fundamental concepts and schemata cannot appear in any possible experiments, and that they therefore play the role of unobservable 'hidden' parameters in the construction of the theory. Only concepts that are contained in a theory of observable quantities are empirically realizable.[1]

In the following investigations the changes of the methodological framework of physics that are based on a theory of the real measuring process and the related changes of the *a priori* structures will be considered in detail. In every case the point of departure will be the critique, expressed by physics, of concepts and structures that in the past have been regarded as *a priori* valid. In particular the structure of space and time, Euclidean geometry, the concept of substance, the law of causality and logic will be examined. In order to be able to discuss meaningfully the critique of these conceptions by physics, it is necessary to set forth the reasons that validate these structures independently of experience. The idea that there are results of science that are applicable to experience even though they do not themselves derive from experience can be traced far back into the history of the exact sciences. Thus, for example, Greek geometry, familiar from the exposition of Euclid, and the syllogistic of Aristotle, are theories whose propositions can be derived from a few axioms, which in turn are valid because of their [internal] evidence. We shall therefore have to inquire into the most important arguments for the validity of *a priori* structures in experience that have been advanced in the history of the exact sciences. In the discussion of space-time structure, the law of causality and the concept of substance, it will be necessary to refer to the arguments of Kant in particular. The reason for this is that Kant, with due regard for all empiricist counterarguments, has given proofs for the validity of the respective *a priori* structures in experience that are relevant for all of modern physics before Einstein, the so-called classical physics. For the problem of the foundation of logic we shall refer, in particular, to the operative foundation of logic, since by this means the validity of the propositions of logic can be demonstrated for all propositions of classical physics.

In order to be able to elucidate the necessity of changing the methodological framework of physics by a theory of the real measuring process it is, furthermore, necessary briefly to review the physical [experimental] results of modern physics. This will be done in such a way that only those points relevant to our philosophical inquiry will be mentioned.

We shall begin the investigation with a discussion of the problems that were raised by the special theory of relativity: questions relating to the structure of space and time, especially the problem of the temporal sequence of events (Chapter I). The special theory of relativity is of particular interest to our problem, for in it Einstein applied for the first time a manner of thinking that has become decisive for all modern physics: the special theory of relativity is a theory of observable quantities that explicitly refers to the possibility of observation and the process of measurement. The physical and mathematical problems of the special theory of relativity are comparatively uncomplicated so that this theory is especially suitable for a discussion of the new epistemological situation. Subsequently problems are considered that were raised by the general theory of relativity, and which question the validity and applicability of Euclidean geometry to empirical space (Chapter II). Apart from the purely mathematical structure of the physical theory, methodological and conceptual questions are here of greater complexity than in the special theory of relativity. Besides the novel structure, by which a theory of observable quantities is characterized, we will for the first time encounter an alternative formulation of theory that operates with hidden parameters, and makes a new form of *a priori* doctrine methodologically possible.

Following this, problems will be discussed that have been posed by quantum theory. The physical results, and in particular the theory of the measuring process in quantum mechanics, will be considered (Chapter III). After this, criticism of the concept of substance (Chapter IV) and of the law of causality (Chapter V) in the quantum theory will be discussed. The last chapter (Chapter VI) deals with the validity and applicability of classical logic for the domain of quantum-theoretical propositions.

NOTE

[1]   P. Mittelstaedt, 'Verborgene Parameter und beobachtbare Größen in physikalischen Theorien', *Phil. Natural.* **10** (1968).

CHAPTER I

SPACE AND TIME

In the analysis of the concepts of space and time in the theory of relativity, a structure has emerged for the first time which is characteristics of all modern physics: the measuring instruments used to determine the properties of nature are themselves objects of that nature. The influence which the laws governing the measuring apparatus exert on the results of measurement must be inherent in these laws, since they are at the same time the laws of the results of the measurement.

In Einstein's theory of relativity this idea was applied consistently to the measurement of space and time intervals by light signals. The concepts of space and time thereby underwent a reinterpretation by which essential properties, attributed to these concepts in Newtonian mechanics, were lost.

The justification for this reinterpretation by reference to experimental results is challenged, however, by the proposition from the philosophy of Kant that space and time as conditions for the possibility of experience cannot, in their structure, depend on these experiences. This objection gives rise to the question of how one can arrive at an interpretation of the results of the special theory of relativity that is both physically and philosophically satisfactory.

1. FORMULATION OF THE PROBLEM

The philosophical aspects of the theory of relativity are of importance for several reasons. For the first time in modern physics a significance is imparted to an approach which, in Kantian terminology, could be denoted as a reflection on the physical conditions of the possibility of experience. These conditions are the laws of nature which govern the measuring process. This new manner of thinking introduced by Einstein, has become of fundamental significance for all of modern physics. As will be shown, it leads to a change in the methodological framework within which physical theories are possible and thereby, ultimately, to a change in the concept of physics itself.

Related to this is the modification in certain methodological assumptions of physics, which, since the philosophy of Kant, have been considered to be unalterable conditions for the possibility of experience. Hence the apparent conflict between the properties of Kant's forms of intuition, space and time, assumed to be *a priori*, and the insights concerning these concepts derived from the theory of relativity, has initiated a discussion of the philosophical situation in modern physics.

A satisfactory discussion of this difficulty is possible only if one does justice to the philosophical as well as the physical side of the problem. To begin with, it should be emphasized that from the philosophical point of view the transcendental character of space and time, and the associated *a priori* judgments of the structure of experience, cannot be corrected by a specific experience. Thus the philosophical objection to the theory of relativity must be entirely respected.

On the other hand, however, the theory of relativity clearly is not a theory whose propositions simply relate to experimental data. Rather, the weight of this theory is in reflecting on the physical preconditions of experience. This reflection, it will be shown, makes it possible for the theory of relativity, in turn, to offer a critique of the transcendental propositions of the philosophy of Kant.

In the following treatment of this complex of problems the physical aspect, that is, the novel structure of the theory of relativity as compared with classical physics, will be considered first. Following this, arguments which were put forth in the philosophy of Kant to justify the *a priori* structure of space and time will be discussed. Confronting these arguments with the results of the theory of relativity will then demonstrate in what sense the theory may be said to cause a revision of the concepts of space and time.

2. THE SPECIAL THEORY OF RELATIVITY

In his *Philosophiae naturalis principia mathematica* (1687) Newton introduced the concepts of absolute space and time as follows: "Absolute space, in its own nature, without relation to anything external, remains always similar and immovable", and "absolute, true, and mathematical time, of itself, and from its own nature, flows equably without relation to anything external".[1] In order that these concepts have a physically verifiable meaning, a measuring process, by which they can be experimentally determined, must be specified.

If one assumes the existence of rigid bodies, then the concept of absolute time acquires a physically realizable meaning. For with their aid it would always be possible to synchronize the clocks of different observers, even though they be in relative motion or separated in space. The time derived from a comparison of clocks could be designated as absolute time, and the time-dependence of all processes could then be related to this base. Identical results would, of course, be obtained if the synchronization were carried out with light signals, provided that light propagates rectilinearly and instantaneously in space.

Defining the concept of absolute space proves considerably more difficult. Measurements of position always involve the measurement of distances and angles in relation to a basis. It will be assumed here as well that the measurement can be executed with a rigid standard. In general a basis to which a set of positions can be referred will be called a reference system. Mathematically such a system corresponds to a definite coordinate system. To begin with, the measurement of positional coordinates provides only relative data which relate to a specific reference system and which therefore represent only a relative space.

It might be possible, however, to choose a particular reference system and, thereby, to define absolute space in which all motion occurs. One could arbitrarily select a reference system, of course, and contend that it represents absolute space. But such a proposition would have no verifiable content. For it turns out that in all relatively stationary or uniformly moving frames of reference, mechanical processes proceed in an equivalent manner. Consequently the stationary state cannot, be experimentally distinguished by mechanical means, from that of uniform motion.

While it is possible for all observers to state, on the basis of the synchronization of clocks mentioned earlier, whether two events, even though they occurred at spatially separated locations, took place at the same time, an analogous procedure is not possible for space. The assertion that two events, separated in time, occurred at the same location, while valid for all stationary observers, will bring divergent conclusions from non-stationary observers.

It is apparent from these comments that, in general, the communication of location data must specify the reference system on which the measurements are based. The reference system represents the experimental preconditions under which the measurements are carried out. In the transition from one reference system to another the measurement data must be transformed according to that transformation which maps the coordinates of one basis into those of the other. A covariant formulation of a theory is one in which the formulation of physical data explicitly contains the experimental preconditions of the measurement in such a way that a change in the preconditions (by a transformation) causes a corresponding change in the physical results. Furthermore, if the fundamental equations are entirely independent of the reference system, as in the case of classical mechanics, then they are called invariant. Covariance and invariance always refer to a specific set of transformations. Transformations which describe transitions to reference systems in uniform motion are denoted Galilei-transformations. Hence classical mechanics is a theory that is Galilei-invariant.

If one designates the coordinates of a reference system K by x, y, z, t, those of K', which is moving in the x-direction, by x', y', z', t', the components of the relative velocities by v_x, v_y, v_z, and if K and K' coincide at $t = 0$, then, for $v_y = v_z = 0$, the Galilei-transformation which leads from K to K', is:

$$x' = x - v_x t \quad y' = y \quad z' = z \quad t' = t.$$

In order to assign a meaning to the concept of absolute space, the Galilei-invariance of classical mechanics notwithstanding, Newton resorted to the inertial effect of accelerated motion. For the experimental realization of absolute space, he proposed two dynamic experiments, with the aid of which the magnitude and direction of motion referred to absolute space could supposedly be measured.[2] One of these is the bucket-experiment, the other involves two globes connected by a string and rotated about their centre of gravity. From the curvature of the water surface and the tension of the string, respectively, motion in absolute space could then be deduced. The discussion which followed, in which Leibniz[3] and Huygens[4] participated decisively, did not find a certain conclusion until Mach[5] interpreted rotation as relative motion with respect to the fixed stars and thereby, finally, deprived absolute space of any mechanical justification.[6]

All the more important was the question whether one could demonstrate motion relative to absolute space by other than mechanical means. For this reason, electromagnetic processes were of particular interest, since electrodynamics was developed completely independently of mechanics. Hence Drude and Abraham posed the question whether it was possible that the ether, carrier of the newly discovered electromagnetic waves, could be interpreted as 'absolute space', thereby giving the concept a meaning, but now from the point of view of electrodynamics. As before, the simplest form of the question involves, first of all, the experimental comparison of uniformly moving systems in the absence of inertial effects, i.e. the so-called inertial systems. Among the many experiments that dealt with this problem, the experiment of Michelson, Morley and Miller obtained a particular fame. In view of the great velocity of light waves, it appeared expedient to use inertial systems also moving at great velocity. In the Michelson experiment, therefore, the rotation of the earth around the sun, which takes place with a velocity of 30 km s^{-1}, was used. This, as well as all other experiments which dealt with this problem, demonstrated that electromagnetic processes proceed in all inertial systems in an entirely identical manner. Experimentally the situation is, therefore, quite similar to that of classical mechanics. The independence of phenomena from the inertial system is called the principle of relativity.

Immediately, however, the following difficulty arises. Apart from the principle of relativity, the principle of the constancy of the velocity of light is known to be valid empirically: Regardless of whether an observer moves towards or away from a light source, the light emitted will travel past him at a velocity of 300 000 km s^{-1}. Therefore, the propagation of light rays is described by the equation:

$$(x - x_0)^2 + (y - y_0)^2 + (z - z_0)^2 = c^2 (t - t_0)^2$$

which is independent of the inertial system chosen, and where x_0, y_0, z_0 are the coordinates of the light source and t_0 is the instant at which the light is emitted. If, however, one uses a Galilei-transformation to map from an

x-coordinate system to an x'-system, then the equation for light propagation does not remain invariant.

The incompatibility of the constancy of the velocity of light and the principle of relativity as expressed by the Galilei-invariance can be illustrated by the following paradox: Suppose a light source which moves with a velocity v relative to observer A is stationary relative to observer B. Then both observers simultaneously see, as wavefronts of the light, spheres the centers of which are at rest relative to them, i.e. they see different spheres. The contradiction disappears when we note that space points that are simultaneously traversed by light for A are not simultaneously traversed for B. The following considerations will show that the solution to the paradox is, in fact, given thereby.

It can be shown that, as a consequence of the equivalence of all points in space and time, transformations that map from one inertial system to another must always be linear functions of the coordinates x, y, z and t. In order to take the principle of relativity into account, the transformation that maps from one inertial system to another must have the form

$$x' = \frac{x - vt}{\sqrt{1 - v^2/v_\infty^2}}, \quad y' = y, \quad z' = z, \quad t' = \frac{t - \dfrac{vx}{v_\infty^2}}{\sqrt{1 - v^2/v_\infty^2}}$$

if one considers two inertial systems translating in the x-direction with a relative velocity v. The quantity v_∞^2 is a still arbitrary constant, which is independent of x, t and v. For the special value $v_\infty = \infty$ one obviously obtains the Galilei-transformation. If one requires, that the transformation also satisfies the principle of constancy of the velocity of light, i.e. the equation for light propagation is left invariant, then it follows $v_\infty^2 = c^2$. Therefore for two inertial systems moving in the x-direction the transformation is

$$x' = \frac{x - vt}{\sqrt{1 - v^2/c^2}}, \quad y' = y, \quad z' = z, \quad t' = \frac{t - \dfrac{vx}{c^2}}{\sqrt{1 - v^2/c^2}}$$

This transformation (given here for a special case) is called the Lorentz transformation. Hence the situation is as follows: Whereas in classical mechanics one can perform the transition from one inertial system to another with the aid of the Galilei-transformation ($v_\infty = \infty$), it is apparent that for the processes of light propagation or, in general, of electrodynamics, this transition is described by the Lorentz-transformation ($v_\infty =$ c). One has to decide, therefore, which of these transformations is the physically correct one. The Lorentz-transformation receives support from the fact that electromagnetic processes are known to a much greater accuracy than mechanical processes, and also by the observation that in the limit, as $c \to \infty$, it approaches the Galilei-transformation. Hence it could be that mechanics in its classical form represents an idealization of the real circumstances, which is proper only in the limiting case $c \to \infty$, and which is subject to certain corrections. The only difficulty with such a solution is the puzzling nature of the Lorentz-transformation, and especially the requirement that, in going from one inertial system to another, time must be transformed as well. In contrast, the use of the Galilei-transformation for such a transition is immediately plausible, at least as long as one starts with the concept of time previously introduced.

At this point Einstein's analysis of the concept of time enters. It starts with the assumption that the concept of absolute time employed in classical mechanics is not a useful one. The method of synchronizing clocks with the aid of effects which propagate instantaneously is not realizable in a material world. The reason for this is that ideally rigid bodies do not exist and that the fastest signal known, light, propagates at a finite velocity. It is therefore quite inappropriate to maintain the concept of simultaneity used in classical mechanics. In its stead Einstein puts a definition which will be denoted as relative simultaneity and which, from the first, takes account of the material constraints.

For events that occur at coincident spatial points the classical concept of simultaneity presents no difficulty and will therefore be retained. However, for events A and B occurring at different points in space, x_A and x_B, which are relatively at rest, the concept of simultaneity will be defined in the following way: Suppose a light signal, emitted from x_A at time t_1, is reflected at x_B and returns to x_A at time t_2 (Figure 1). Then it will be said that clocks at x_A and x_B are synchronized if at the moment of the light reflection the time indicated at x_B is $(t_1 + t_2)/2$.

The application of this concept to concrete events apparently poses no difficulties. That it is a plausible definition of simultaneity is seen when one notes that the velocity of light is the same in both directions. It follows immediately that if clock B is synchronous to A then clock A is also synchronous to clock B, and hence that simultaneity is a symmetrical property. Furthermore, it is transitive: If A is synchronous to B, and B is

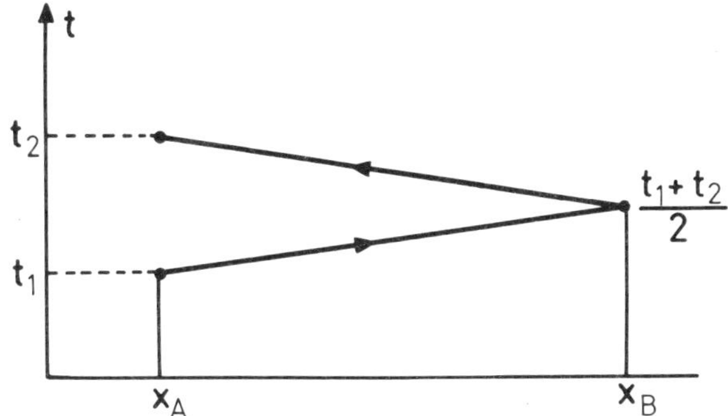

Fig. 1. Schematic representation of Einstein's definition of simultaneity.

synchronous to C, then A is synchronous to C. If one introduces a two-place function $S(x, y)$ for synchronicity, then the following relations hold:

$$S(A, A)$$
$$S(A, B) \rightarrow S(B, A)$$
$$S(A, B), S(B, C) \rightarrow S(A, C)$$

The concept of relative simultaneity therefore satisfies the laws of an equality relation and hence is formally acceptable as well.

Choosing to use light signals for the definition of simultaneity without searching for other, possibly faster, methods of signal transmission may at first appear arbitrary. The following considerations will show, however, that the velocity of light can be regarded as a limiting velocity which cannot be exceeded by the velocity of any other possible signal. To undertake the synchronization of clocks by light signals is therefore entirely reasonable.

For moving reference systems the use of a concept of simultaneity that is based on light measurements leads to results which deviate from those derived from traditional concepts. An example introduced by Einstein, a train moving with respect to the same track, illustrates this. Suppose two points on the track, x_A and x_B, are separated by a distance $2s$ (Figure 2). Let the centre of this distance be at x_M. Suppose the center point, x'_M, of a train which is moving with a velocity v relative to the track coincides at $t = 0$ with x_M. Further, at $t = 0$ let light flashes be emitted at x_A and x_B, which arrive at x_M simultaneously at times $t_A = t_B = s/c$. This is the situation that an observer at x_M will perceive. Consequently he will state that the light flashes occurred simultaneously.

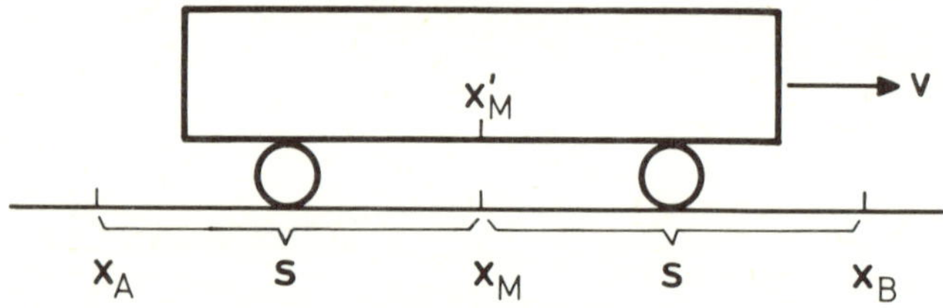

Fig. 2. Simultaneity in moving systems.

At x'_M, however, the two signals do not arrive at the same time (Figure 3). Since the train is moving with a velocity v from x_A towards x_B, the signal from x_B reaches it sooner than that from x_A, and indeed the signals arrive at the times

$$t_B^{(v)} = \frac{t_B}{1 + \dfrac{v}{c}}, \qquad t_A^{(v)} = \frac{t_A}{1 - \dfrac{v}{c}}$$

at x'_M, that is, at the points $x_A^{(v)} = x_M + v t_A^{(v)}$ and $x_B^{(v)} = x_M + v t_B^{(v)}$. (It is convenient to let $x_M = 0$). This is still the description that an observer at rest at x_M would give for the process. It is for the time being quite

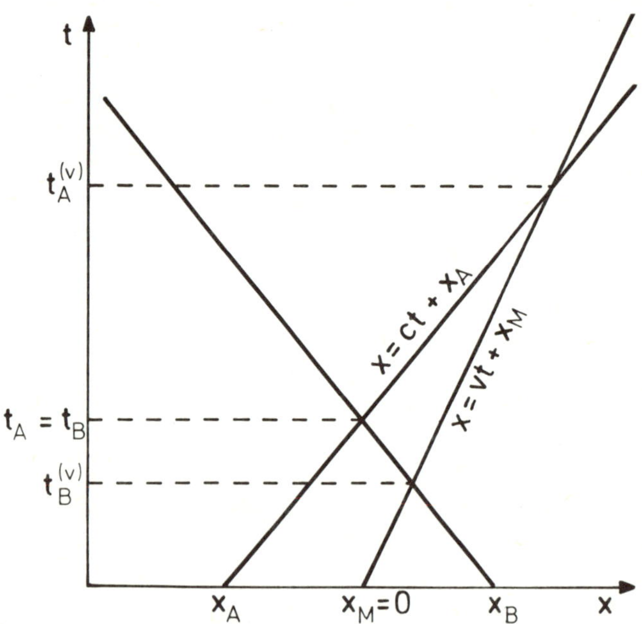

Fig. 3. Simultaneity in moving systems.

unproblematic, for the observer at rest at x_M could of course maintain that the two flashes of light are simultaneous *per se*, and that they only appear separated by a time interval

$$\Delta t^{(v)} = t_B^{(v)} - t_A^{(v)}$$

to the moving observer, precisely because he is moving.

The situation is altered, however, if we proceed to ask what an observer on the train, that is, an observer at rest with respect to x_M', perceives. In this case the two events, the arrival of the light signals at x_M must be submitted to a Lorentz-transformation. If we denote the times at which the moving observer registers the arrival of the light signals at x_M by $t_B^{(v)'}$ and $t_A^{(v)'}$ respectively, then it follows that

$$t_B^{(v)'} = t_B^{(v)} \sqrt{1 - \frac{v^2}{c^2}}, \quad t_A^{(v)'} = t_A^{(v)} \sqrt{1 - \frac{v^2}{c^2}}$$

and that the time difference between the signals, $\Delta t^{(v)'}$, is

$$\Delta t^{(v)'} = t_B^{(v)'} - t_A^{(v)'} = -\frac{2sv}{c^2} \bigg/ \sqrt{1 - \frac{v^2}{c^2}}$$

The observer stationary at x_M' can now, with equal justification, maintain that $\Delta t^{(v)'}$ is the actual time interval between the light flashes. Because of the invariance of the equation of light propagation, it is true for him, as well as for the observer at rest at x_M, that the wavefronts of the two light signals form spheres at every instant. It is therefore meaningless to say that one observer is at rest whereas the other is moving. For both, the situation is entirely equivalent.

It appears that it is impossible to give the concept of the simultaneity of two events a meaning which is valid in all inertial systems. Simultaneity defined by light measurements is therefore appropriately denoted as *relative simultaneity*. For it is always meaningful to say that two events occur simultaneously relative to a certain observer. A moving observer, however, will not state that these events have occurred simultaneously.

These considerations show that transforming the time-coordinates along with the position-coordinates is reasonable and necessary when going from one inertial system to another. In the stationary system of the above example the two flashes of light are separated in space by $\Delta x = 2s$, and in time by $\Delta t = 0$. In the moving system, on the other hand, it proved to be:

$$\Delta t^{(v)'} = - \frac{2sv}{c^2} \bigg/ \sqrt{1 - \frac{v^2}{c^2}}$$

But this is precisely the value which results if one subjects $\Delta x = 2s$, $t = 0$ to a Lorentz-transformation. It thus appears that, in going from one inertial system to another, time intervals, measured by Einstein's light ray method, transform exactly as is required by the Lorentz-transformation. We can therefore regard a time interval, as measured by Einstein's method, as a consistent and materially realizable model for the parameter t appearing in the Lorentz-transformation.

It is important to note, that for the derivation of this consistency property of the Lorentz-transformation the principle of constancy of the velocity of light must be used and that the principle of relativity is not sufficient for this derivation. That is, if one omits the principle of constancy of the velocity of light and uses for the transition between two systems of inertia the generalized Lorentz-transformation with arbitrary v_∞, then the discussion of the train experiment leads to the following result: For the time difference $\Delta t^{(v)'}$ between the signals, which is registered by the moving observer, one obtains

$$\Delta t^{(v)'} = - \frac{2sv}{c^2} \frac{\sqrt{1 - v^2/v_\infty^2}}{1 - v^2/c^2}$$

whereas the transformation of the values $\Delta x = 2s$, $\Delta t = 0$ yields the difference

$$\Delta t' = - \frac{2sv}{v_\infty^2} \bigg/ \sqrt{1 - \frac{v^2}{v_\infty^2}}$$

Both values agree only if $v_\infty = c$, i.e. if one uses Lorentz-transformations. Therefore the measured time intervals are a consistent model only for that time coordinate, which has been calculated by a Lorentz-transformation.

Hence for the concept of time defined by the measurement of light, the application of the Lorentz-transformation to the description of the transition from one inertial system to another is as fully convincing as was the application of Galilei-transformations under the assumption of absolute time.

Considerations very similar to those used for the concept of time, can also be applied with regard to the measurement of position. The employment of light signals for the measurement of lengths leads in a similar manner to the Lorentz-transformation of the position-coordinates. But since the difference of these formulas from the corresponding Galilei-transformation (except for the very important dependence of length on speed, the so-called Lorentz-contraction) is not of such fundamental significance as the fact that there cannot be an absolute time, a discussion of the details of position measurement will not be given.

After it has been clarified in this way in which sense position and time parameters appearing in the Lorentz-transformation can actually be interpreted as the physical magnitudes of space and time, there is no objection to the general application of the Lorentz-transformation to transitions between inertial systems. Because of the principle of realtivity, the laws of mechanics must also be invariant under Lorentz-transformations.

This can only be achieved by certain corrections of the laws of classical mechanics. However, the laws of relativistic mechanics corrected in this way approach those of classical mechanics in the limit as $c \to \infty$. This can be understood insofar as an infinitely great velocity of light makes instant signal transmission possible and thereby the introduction of absolute time, as is assumed in classical mechanics.

Essential changes in the temporal sequence of events follow from Einstein's notion of time. For if we consider two events (x_1, t_1) and (x_2, t_2), then two possibilities exist: Either

$$c^2 (t_2 - t_1)^2 - (x_2 - x_1)^2 \geqslant 0$$

in which case one calls the events time-like, or

$$c^2 (t_2 - t_1)^2 - (x_2 - x_1)^2 < 0$$

when the events are space-like. The essential difference between the two cases is that the temporal sequence of time-like events is the same in all reference systems, whereas the sequence of space-like events depends on the choice of reference system. This can easily be illustrated in an x-t diagram if one of the events is placed at the origin of the coordinates (Figure 4). If the second event is located in the hatched area — i.e. the so-called light cone — then it is time-like to (x_1, t_1). If it is outside the hatched area it is space-like to (x_1, t_1). The boundary between the two areas corresponds to the propagation of light.

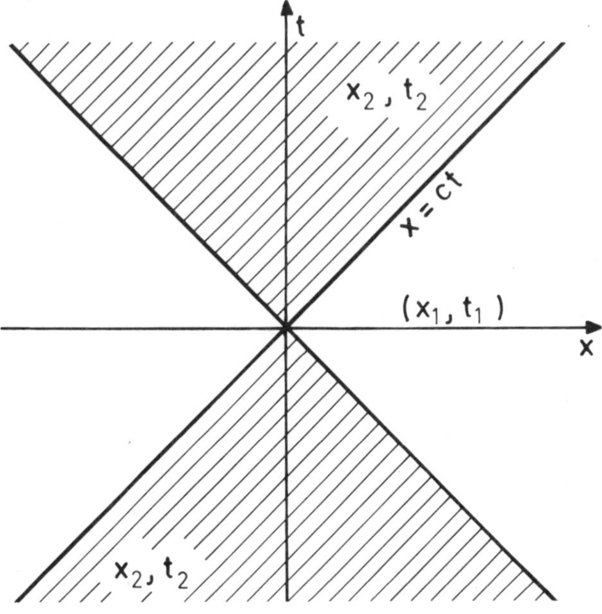

Fig. 4. The light cone.

If for two events one is said to be the cause of the other, then the sequence in time of these events must be independent of the reference system. Events that are causally related must therefore be time-like, i.e. if (x_1, t_1) is the causal event, then the effect event (x_2, t_2) must lie in the part of the light cone that is directed upward, the so-called forward cone. Correspondingly, (x_1, t_1) can only be an effect event, if the causal event (x_2, t_2) lies in the lower part of the light cone, the backward cone.

The world lines of causal processes must therefore be contained within the light cone, that is, they must propagate with a velocity $v < c$. The greatest possible velocity with which effects can propagate is the velocity of light, c. Otherwise the events are space-like and therefore can no longer be considered causally related. The velocity of light thus functions as a limit velocity to the propagation of any effect. But since signals also are causal events, this consideration corroborates the assumption made in the beginning, namely that the best possible synchronization of clocks is by signals travelling at the velocity of light.

These considerations make it possible to transpose by analogy the notions of past, present and future, familiar from classical physics, to the new situation. For this purpose we shall use the following definitions. For a certain event E at the world point (x_1, t_1) the future consists of all events that can be causally affected by E, and the past of all events that could have affected E. Then the present comprises all events that are neither affected by E nor themselves influence E.

Conceptually these definitions agree completely with what in classical physics is called the past, present and future; the mathematical content, however, will be very different depending on whether they are used in conjunction with classical or relativistic physics.

In the x-t diagram of an event (x_1, t_1), space-time is partitioned as shown in Figure 5.

The upper part of the light cone corresponds to the future, the lower to the past. All other points, which lie in space-like relation to (x_1, t_1), represent the present. The last concept, especially, differs formally from the concept of the present in classical physics. There the present includes only those points which lie on the x-axis, whereas in this case the entire x-t domain of space-like points delineated by the light cone must be included.

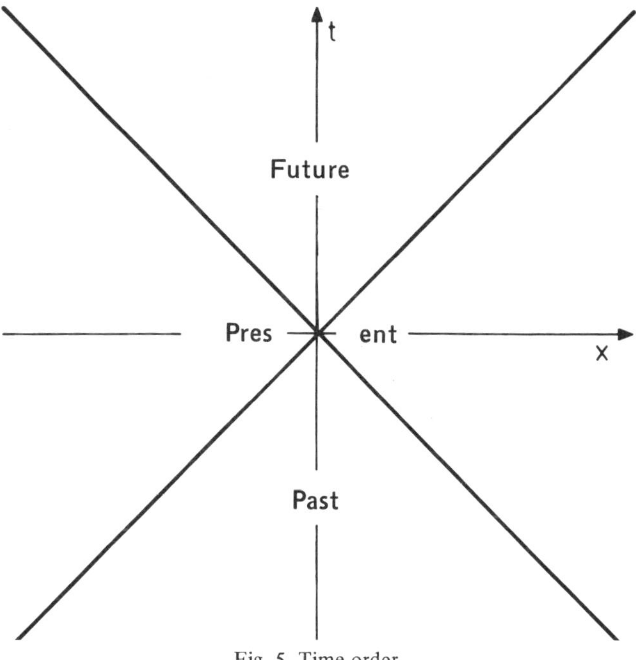

Fig. 5. Time order.

3. THE STRUCTURE OF PHYSICS IN THE THEORY OF RELATIVITY

The relations between the concepts of space and time, the physical realization of these concepts, and the transformational properties of physical theory, were discussed in the last section with reference to both classical and relativistic physics. In the course of this, it was shown that the physical conditions under which measurements of space and time intervals are made determine certain properties of the experimental data obtained.

The principle of relativity of classical mechanics is an empirical conclusion. If one starts with the relevant presupposition of prerelativistic physics, that lengths and time intervals can be measured with rigid bodies or with an instantaneous signal (e.g. light), then this assumption together with the principle of relativity has the consequence that classical mechanics must be a Galilei-invariant theory.

Subsequent considerations have demonstrated that in nature neither ideally rigid bodies nor infinitely fast light exists. The preconditions of classical theory, namely that space and time measurements can be carried out with such standards, are therefore not realizable in the material world. Thus, the requirement arises of carrying out fundamental measurements with the aid of such objects and processes as actually occur in nature. The development of the theory of relativity has shown that the medium best suited for the measurement of space and time coordinates is light.

The laws dealing with light are determined by electrodynamics. But for electrodynamics, as well as for mechanics, a principle of relativity has empirical validity with respect to reference systems in uniform motion. In connection with the use of light for the measurement of position and time, this has the consequence that the theories of electrodynamics and mechanics, that is, all of macroscopic physics, must be Lorentz-invariant rather than Galilei-invariant.

By associating the measurement of space and time with the laws of light, the theory acquires a structure entirely novel compared with that of classical mechanics: the means by which nature is perceived, are nothing other than parts of this very nature. The physical laws are at the same time the laws of the measuring instruments and hence the physical conditions under which experimental results can be obtained at all. The influence which these laws exert on the structure of the results of measurements must already be contained in the laws, since they are at the same time valid for the measuring results themselves.

In the last section this was precisely demonstrated for time measurements in different inertial systems. The laws of the propagation of light in moving systems are the physical preconditions under which measurements of time must be undertaken. The influence that these conditions have on the time intervals actually measured must already be contained in the laws of light propagation. This finds expression insofar as the measured time coordinates, in going over to another inertial system, transform according to the Lorentz-transformation. The measurement of the space-time coordinates with the aid of light therefore represents a realizable and consistent model for the variables x and t that appear in the equation of propagation of light.

In discussing classical mechanics, we have interpreted the concept of absolute time, on which this theory is based, with the aid of measurement processes which presuppose the existence of rigid bodies or instantaneous light signals. It must be emphasized, however, that already in this exposition a manner of thinking is employed which indeed was first introduced into physics by Einstein. Reflection on the material conditions under which an experimental realization of the concepts of space and time would be possible, was at no place carried out by the founders of classical mechanics. Surely such a reflection would have called forth certain doubts about the fundamental concepts of mechanics; for at no time was there a plausible indication of the existence of rigid bodies, and the fact that light does not propagate instantaneously was also experimentally certified by Olaf Römer in his measurement of the velocity of light.

The reason that classical mechanics was never questioned is that, before Einstein, the need of defining physical concepts by stating the measurement process was never even recognized. Physics was understood as the science of a nature that was quite independent of the observer; a science which could be experimentally tested, but which was not affected in its structure by this possibility. This conception of physics presented no difficulties as long as the influence of the measurement process on the results was still negligibly small compared to the errors which appeared independently of the measurement process. Hence, for a long time, it was certainly of no consequence to mechanics that the velocity of light is only 300 000 km s^{-1} and not infinite, since almost all velocities appearing in mechanics could be regarded as very small compared to this magnitude.

The situation changed, however, when the techniques of observation, especially those of electromagnetic processes, were refined to the point where the real properties of the measuring instruments could no longer be disregarded. The laws of the measuring instruments (that is, the wave equation of light) thereby exerted a direct influence on the results of the measurements. These could therefore no longer be interpreted as statements about unobserved nature, since they always contained features that could be traced back to the measuring process.

As a consequence, physics was confronted by alternatives with regard to its aims: *Either* physics is a science of nature *per se* without regard to the possibility of observation. Then the classical concepts of space and time can be retained, but the propositions of theory can no longer be experimentally verified. A confrontation with experiment could then only be achieved if a theory of the real measuring process were added to physics. *Or* else physics describes nature as it reveals itself when it is investigated with real standards and clocks. Then it must accommodate its concepts to this new situation and thereby become a theory, the assertions of which can all be related to experimentally realizable processes. In prerelativistic physics, such a decision as to the aims of physics did not have to be made, for there both possible aims still coincided. In modern physics, however, they are mutually exclusive.

The decision as to which path was to be pursued for physics took place clearly with regard to the goal of being able to describe experimental results and to eliminate all unobservable magnitudes from theory. Since Einstein, therefore, physics is a science which is expressly understood as a theory which describes nature as it appears when investigated with real measuring standards and clocks.

4. SPACE AND TIME IN THE PHILOSOPHY OF KANT

In contrast to the theory of relativity, from the outlook of philosophical apriorism, and according to the meaning of the Kantian philosophy, space and time must be counted among the transcendental conditions for the possibility of experience, and therefore that a change in the concepts space and time can never occur from this experience.

This objection is directed primarily against Einstein's relativity of time, which appears to be contrary to the unity of time in the transcendental approach. By 'unity of time' in the sense of Kant is meant the assertion that all empirical events are ordered in a uniform continuum of time, and that the sequence of the events is established objectively and independently of the reference system. This unity of time according to Kant exists, with necessity for all objects of experience. In the theory of relativity, on the other hand, the sequence of space-like events always depends on the reference system of the observer.

In the philosophical discussion, the changes in the concept of time in relativity theory diminish the importance of the dependence of spatial dimensions on the reference system. In fact, the changes in spatial dimensions are considerably more subtle and mainly of a more quantitative nature as compared with those of the concept of time. The question, whether position coordinates transform in accordance with the Galilei-transformation as $x' = x - vt$, or the Lorentz-transformation,

$$x' = x - vt \left/ \sqrt{1 - \frac{v^2}{c^2}} \right. ,$$

is probably of lesser interest for philosophical discussion, since it hardly appears possible — from the epistemology of Kant, for example — that a convincing position for either of the transformation formulas can be justified; similarly, change of the time scale (with retention of the time order), occurs in the Lorentz-transformation of time-like events. And as for the Lorentz-contraction, this time-dilatation is also too quantitative for any reasonable comparison with philosophical investigations which may be carried out.

For these reasons, the following discussion will be concerned exclusively with the problem of the uniformity of time in the sense of a uniformly fixed time sequence of empirical events.

In the philosophy of Kant, the necessity for a uniform order of time is based on a precise analysis of the cognitive process. First of all, it is pointed out that knowledge of objects is always associated with intuition: "In whatever manner and by whatever means a mode of knowledge may relate to objects, intuition is that through which it is in immediate relation to them, and from which all thought gains its material."[2] For the object of an empirical intuition, matter and form are distinguished, wherein form is "that which so determines the manifold of appearance that it allows of being ordered in certain relations".[8] It is essential for the transcendental approach that the form of phenomena, as the condition under which phenomena are at all possible, is itself not to be regarded as a phenomenon, but rather "must lie ready for the sensations *a priori* in the mind".[9]

Investigation of this pure form of sensuous intuition had the consequence that there are exactly two such forms, namely space and time. These two forms of perception should therefore "lie ready for the sensations *a priori* in the mind" so that all phenomena actually accessible to us are phenomena in space and time.

In the metaphysical discussion of the concepts of space and time, it is pointed out that space and time are not empirical concepts that can be acquired by abstraction from experiences, since experiences only become possible by the conception of space and time. Since "we can never represent to ourselves the absence of space"[10], and phenomena are not situated in time, then space and time are necessary conceptions which are at the basis of all phenomena. Space and time are therefore to be considered as the conditions for the possibility of phenomena.

The fact that space and time are not discursive concepts but rather pure intuitions has the consequence that, based on this ever present intuition, certain propositions about the structure of phenomena are *a priori* valid. Kant included among the synthetic *a priori* judgments which can be deduced in this manner from the intuition of space, all the axioms of Euclidean geometry.[11] From the form of intuition of time, it should follow that time is one-dimensional and that as a consequence different times are not simultaneous but rather sequential.

The interpretation of space and time as conditions for the possibility of phenomena reveals two properties of these concepts which, at first, appear to be mutually exclusive. The fact that space and time themselves are not objects, but only conditions by which the experience of objects is made possible, constitutes the 'transcendental ideality' of these concepts. But on the other hand, these concepts also possess an 'empirical reality'. For all things that we experience are in space and time, and the structure which space and time have as pure forms of intuition is therefore also present in empirical objects. "Time is therefore a purely subjective condition of our . . . intuition . . . and in itself, apart from the subject, is nothing. Nevertheless, in respect of all appearances, and therefore of all the things which can enter into our experience, it is necessarily objective."[12]

The assertion of the uniformity of time refers to the temporal arrangement of objects that appear in experience. In order to be able to discuss this assertion, it must first of all be clarified how space-time perceptions must be combined to become objects that can occur in experience. Furthermore it must be clarified how a temporal sequence can be established for empirical events. This is of importance since time *per se* is not an object of perception: "Time cannot be perceived in itself, and what precedes and what follows cannot, therefore, by relation to it, be empirically determined in the object."[13] Hence it is still necessary to establish precisely when it is possible to say of two events that one is earlier and the other later. As Kant demonstrated in his 'Theory of the Schematism of the Pure Concepts of Understanding', these two problems are very closely related. The ordering of phenomena to the objects of experience has to take place with certain concepts of understanding, the so-called categories. But the forming of objects of experience through categories comes about precisely because the manifold quality of perceptions is ordered according to time relations. The structure of the time-order of empirical events that comes about in this manner is therefore determined by the categories which were employed for the ordering of perceptions.

Hence the 'uniformity of time' can be a property that is necessarily appropriate for all empirical events only if the temporal ordering, which is produced by the application of categories to phenomena, has this property *a priori*. To what extent this is the case will be demonstrated by the following investigation.

The categories as pure concepts of the understanding can gather and order the phenomena of space-time into objects of experience only when it is established "how pure concepts can be applicable to appearances".[14] Such an application is only possible when a mediating conception exists, which is "intellectual in one respect, sensuous in another". This mediating conception Kant calls the transcendental schema. This schema of pure concepts of the understanding proves to be the transcendental determination of time. Further investigation then demonstrates that the rules that determine the time-order, which is of special interest here, are given by the schemata of the categories of relation, i.e. substance, causality and interaction.

How the temporal sequence of concrete events is actually to occur, is expounded in Kant's 'Analogies of Experience'. The ordering of phenomena with the aid of the category of causality establishes which event is earlier and which is later. If two phenomena can be considered as cause and effect of a process, then the temporal sequence is thereby given.

Now since absolute time is not an object of perception, this [determination of position in time] cannot be derived from the relation of appearances to it. On the contrary, the appearances must determine for one another their position in time ...[15]

Hence the temporal ordering of phenomena with the aid of the category of causality makes it possible to state of the times t_A and t_B of two events, A and B, whether $t_A \leqslant t_B$ or $t_B \leqslant t_A$. From this, the introduction of a definition of simultaneity follows directly: the two events A and B occur simultaneously when both $t_A \leqslant t_B$ and $t_B \leqslant t_A$. From this it is obvious that $t_A = t_B$.

This method of defining simultaneity is then actually used by Kant in the 'Third Analogy of Experience'. The events A and B are simultaneous if A is the cause of B and vice versa. Since time itself is not an object of perception,

there must be something through which A determines for B its position in time, and also reversewise B for A, because only on this condition can these substances be empirically represented as coexisting. Now that alone which can determine the position of anything else in time is its cause ... Each substance ... must therefore contain in itself the causality of certain determinations in the other substance, and at the same time the effects of the causality of that other.[16]

It is clear from these comments how, by applying the categories of relation to phenomena, these can be brought into temporal order. Moreover, as will be shown in detail in the next section, these categories are chosen such that, with their aid, events are actually brought into a uniform temporal sequence which is objectively established independently of the reference system of the observer.

The application of the transcendental scheme to perceptions brings about what Kant calls the uniformity of time.

This unity of time-determination is altogether dynamical. For time is not viewed as that wherein experience immediately determines position for every existence. Such determination is impossible, inasmuch as absolute time is not an object of perception with which appearances could be confronted. What determines for each appearance its position in time is the rule of the understanding through which alone the existence of appearances can acquire synthetic unity as regards relations of time; and that rule consequently determines the position (in a manner that is) *a priori* and valid for each and every time.[17]

Whenever phenomena can be ordered into objects of perception, by the categories mentioned, these will be arranged in a uniform temporal sequence.

5. CRITIQUE OF THE CONCEPT OF TIME
IN THE THEORY OF RELATIVITY

In Kant's presentation of the process of cognition, categories are specified with the aid of which phenomena can be arranged into a temporal order: the causal relation between two events determines their temporal sequence and the reciprocal causal influence correspondingly establishes their simultaneity. Kant, it is true, only intimates how this temporal ordering can be realized in concrete instances. But it is clear from the concept of simultaneity that Kant uses that for the propagation of effects which define simultaneity, in any case, no upper limit to velocity is fixed. Since the causal relation between two simultaneous events must be reciprocal, the definition of simultaneity would otherwise be applicable only to events that occurred at the same place. But, as is apparent from the 'Third Analogy of Experience', this is not at all the intention. It is explicitly stated that this definition is applicable to 'objects widely separated'. But this is only the case if one assumes the possibility of instantaneous propagation of effects (by rigid bodies or light). The definition of simultaneity then permits synchronization of all clocks in the sense of an absolute time, so that the time ordering of all events can be conducted in a uniform manner for all reference systems. Hence the use of Kantian categories has the consequence that all empirical events are arranged in uniform time.

A physical theory must consider the time dependence of empirical events. The theory that accomplished this in complete harmony with the Kantian conceptions is classical mechanics. In it the uniformity of time is secured by absolute time, which is the same in all inertial systems. For an experimental realization of absolute time the possibility of instantaneous signal transmission must be assumed here as well.

As in classical mechanics, the theory of relativity concerns itself with the time dependence of empirical events. The essential difference however, is, that in the theory of relativity it is demanded that all measurements can be performed with real, existing measuring instruments. As is then shown subsequently in this theory, the most suitable standard for the conduct of space-time measurements is light. This, however, does not propagate instantaneously but with a finite velocity.

The fact that the velocity of light is finite, and the fact, derived in the theory of relativity, that causal effects can only propagate at a velocity

$v \leqslant c$, i.e. within the light cone, has the consequence that the method of transcendental time ordering is only applicable to a very restricted domain of events. This implies that the finiteness of the velocity of light, though not in contradiction to the Kantian concept of time sequence and simultaneity, does nevertheless severely restrict its applicability.

If two events, A and B, are causally related so that A is the cause of B, then B must be in the forward light cone of A. Hence the Kantian definition of time sequence, namely that $t_A \leqslant t_B$ when A is the cause of B, leads to a reasonable result. For the time-like separation of causal events, $T = 1/c \sqrt{c^2 (t_A - t_B)^2 - (x_A - x_B)^2} \geqslant 0$ is invariant, so that the temporal sequence of these events is determined independently of the reference system. However, this definition does not offer a possibility of making an assertion about the time sequence of events that lie outside of the light cone.

The definition that Kant uses for simultaneity can also continue to be used; however, even stricter limits are imposed on its application. If one is to continue defining $t_A = t_B$ by the validity of $t_A \leqslant t_B$ and $t_B \leqslant t_A$, then this implies that $t_A = t_B$ is exactly valid when B is in the forward cone of A and A lies in the forward cone of B. But this is indeed only possible if the space points of the two events coincide, that is, when both events occur at the point (x_A, t_A). The simultaneity $t_A = t_B$ is then valid independently of the reference system. But for all other events that are space-like related to A this criterion of simultaneity is not at all applicable.

The method which Kant uses for ordering time is therefore applicable to all events which lie within the light cone. Under these conditions, one then obtains the same results as one would with the assumption of an infinite velocity of light:

(*a*) The temporal sequence of events is unambiguously determined by their causal relation.

(*b*) The simultaneity of two events is unambiguously determined by their reciprocal causality.

(*c*) There is a uniformity of time insofar as the sequence or simultaneity of all events determined by (a) or (b) is established independently of the chosen reference system.

The disadvantage of this method derives from its restriction of events within the light cone. Especially for the simultaneity definition this is very disadvantageous, for it is not possible ever to make an assertion about the

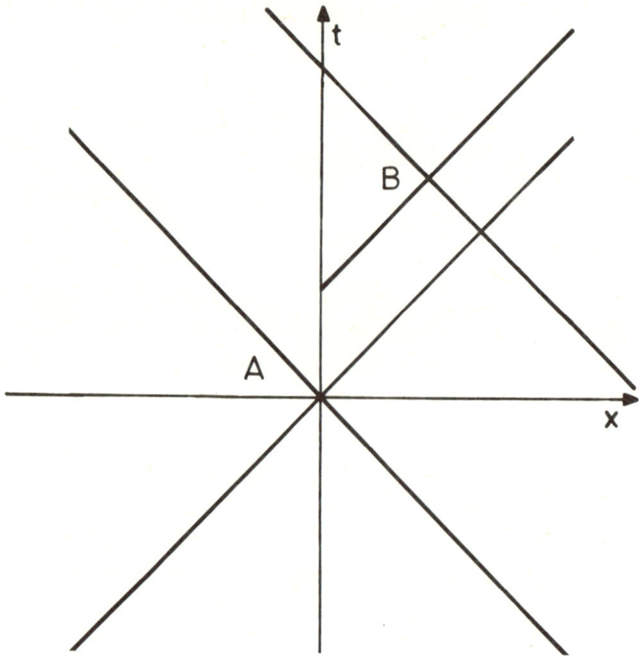

Fig. 6. Time order of two events.

simultaneity of space-like events. But it is immediately apparent that *all* events are covered by these definitions if $c \to \infty$. The light cone is then extended to the *x*-axis, so that all events become time-like related. This confirms our earlier observation that only under the assumption of instantaneous signal transmission can the Kantian categories be applied to all events.

But since the velocity of light is in reality finite, and since therefore the Kantian definitions only have a limited domain of application, Einstein introduced a new definition of simultaneity, one which from the start takes account of the finite velocity of light, yet is applicable to spatially separated events. For spatially coinciding events it agrees with the old definition. Since it can be shown from Einstein's concept of simultaneity that effects propagate with, at most, the velocity of light, the concept of causality also undergoes a change: causal relations can occur only between time-like related events. These new categories are chosen such that they meet the empirical requirements of the possibility of experience: they are applicable to all phenomena, even under the constraint that the measurements be carried out with real standards.

In this connection, Einstein repeatedly emphasized that categories are not given *a priori*, and that the choice of the concept of understanding has to take place in such a way that these concepts may apply to reality:

I am convinced that the philosophers have had a harmful effect upon the progress of scientific thinking in removing certain fundamental concepts from the domain of empiricism, where they are under our control, to the intangible heights of the *a priori.* [18]

It must be admitted that this reproach can hardly be directed at the philosophy of Kant, for nowhere has Kant proposed categories to be necessary for thought. Categories belong to the conditions which make the knowledge of things at all possible. It can quite possibly occur, therefore, that certain perceptions cannot be consolidated with the aid of categories into objects of experience. Kant, to be sure, believes that in that case "no such thing as knowledge would ever arise".[19] In view of the conceptual developments of modern physics, we are unable to follow him in this point. For the new concepts of time order, which were introduced by Einstein, can be applied to perception equally as well, and in some circumstances even better than, those of Kant. Similarly the possibility of finding propositions about experience, which are *a priori* valid, is not restricted. Certain laws of the objects of experience, which · have been formed with the aid of categories from perceptions, are *a priori* valid in the sense of Kant. But there are also similar assertions that are *a priori* valid with Einstein. They are valid for all events which have been interpreted with the help of Einstein's concepts of causality and simultaneity. In general they differ, however, from the propositions that are based on Kantian categories and that are treated in the 'Analogies of Experience'. Only when two events are time-like related, do these assertions agree with the results of Kant cited above (a), (b), (c).

For events that are space-like related, and about which nothing can be said using the old concepts, one obtains essentially different results:

(ā) The temporal sequence of space-like related events is dependent on the reference system used. Causal relations between such events cannot exist.

(b̄) The existence of simultaneities between space-like related events depends on the reference system used. Simultaneous events, insofar as they are space-like related, have no reciprocal interaction.

($\bar{c}$) The uniformity of time and also the possibility that it is objectively reasonable to speak of a temporal sequence of events, are lost. As mentioned in ($\bar{a}$) and ($\bar{b}$), the sequence and simultaneity of space-like related events always depends on the reference system.

The relations between the different categories of Kant and Einstein, their domains of application and the *a priori* valid assertions therein, are summarized in Table I.1 (where E indicates an event and L is the set of events within the light cone). This comparison of Kant's and Einstein's concepts of simultaneity and causality as well as of their assertions about the relation between time order and causality clarifies the following: Even in view of the new situation in the theory of relativity, the relations maintained by Kant in the 'Analogies of Experience' do not prove to be false. For if the temporal ordering of phenomena can be achieved with the concepts of Kant, then (a), (b), (c), our assertions on the relation of causality and temporal sequence and the uniformity of time resulting therefrom, continue to be valid. However, as we have seen, this is only possible for events within the light cone.

For space-like related events the considerations of Kant are not false since these events are not covered by the Kantian categories.

The inclusion of all events would be possible only under the unrealistic assumption that the velocity of light is infinite. For in the transition to the case $c \to \infty$ the light cone is extended to the x-axis, so that the separation of all events becomes time-like. Then it would be possible to establish a time sequence for all events, which would be valid in all reference systems. Assertions about simultaneity would then be independent of the reference system.

TABLE I.1

Concept	Speed of Light	Domain of application	Proposition
Causality Simultaneity according to Kant	$c \to \infty$ $c < \infty$	all E $E \in L$	a, b, c a, b, c
Causality Simultaneity according to Einstein	$c \to \infty$ $c < \infty$	all E $E \in L$ $E \notin L$	a, b, c a, b, c $\bar{a}, \bar{b}, \bar{c}$

One is led to a contradiction between the results of Kant and those of the theory of relativity only if one imposes the time relations that are present in the application of Kantian categories and accordingly are valid in their domain of application (i.e. the interior of the light cone) on all events. An example of this is the experiment mentioned earlier, in which the spherical propagation of a light signal is viewed from a stationary and a moving reference system: Every observer sees the wavefront of a sphere, the center of which is at rest relative to him; hence everyone sees a different sphere.

The inconsistency disappears immediately if, instead of the Kantian concept, one uses Einstein's concept of relative simultaneity, since the latter is applicable to all, even spatially separated, events. Although it is true that it is no longer possible to define a uniform time for all reference systems with this concept, still, it is not possible with any other concept of simultaneity that allows for the finiteness of the velocity of light, either. The usefulness of Einstein's concept for physics consists in its applicability in similar manner to all events. For events within the interior of the light cone, however, results are obtained, which also follow from the old concept of simultaneity.

All these considerations show that the new and in part unusual results of relativity theory, as for example the disappearance of the uniformity of time, can be traced to changes in the very concepts with the aid of which experiences are at all realized.

The changes to which the concepts of causality and simultaneity have been subjected in the theory of relativity, were stimulated by reflection on the physical conditions of the possibility of experience. The Einsteinian demand that scientific concepts be realizable in the material world require the definition of new concepts which are actually applicable to experience. Einstein's concept of simultaneity is of this type.

Whereas the older physics was always to be understood as a theory of nature-in-itself, and as a consequence either never reflected on measurement processes, or contented itself with unrealizable definitions – such as measurement with ideally rigid bodies –, this is all quite different in modern physics since Einstein. The conception of physics was changed insofar as physics is now a theory of nature as it reveals itself when investigated with real measuring rods and clocks. This new concept of physics has become decisive for all further developments. Its significance became apparent for the first time in Einstein's analysis of space-time measurements of phenomena at high velocities.

NOTES

[1]   I. Newton, *Philosophiae naturalis principia mathematica*, 1687; English translation F. Cajori, *Sir Isaac Newton's Mathematical Principles of Natural Philosophy*. University of California Press, Berkeley, 1934, p. 6.

[2]   *Ibid.*, p. 10.

[3]   G. W. Leibniz, Correspondence with Clark. [See P. G. Lucas translation of the *Leibniz-Clerke Correspondence.*]

[4]   Ch. Huygens, *Oeuvres complètes*, Vol. 10 (Correspondence 1691–1695).

[5]   E. Mach, *Die Mechanik in ihrer Entwicklung*, Leipzig 1883. [English translation T. J. McCormack, Open Court, Chicago, n.d.]

[6]   Compare p. 68

[7]   Immanuel Kant, *Kritik der reinen Vernunft*, 1781(A), 1787(B); English edition of N. K. Smith, *Immanuel Kant's Critique of Pure Reason*, Macmillan, London, 1929, p. (B)33, p. 65.

[8]   *Ibid.*, p. (B)34, p. 66.

[9]   *Ibid.*, p. (B)34, p. 66.

[10]   *Ibid.*, p. (B)38, p. 68.

[11]   Compare p. 42

[12]   Kant, p. (B)51, p. 77.

[13]   *Ibid.*, p. (B)233, p. 219.

[14]   *Ibid.*, p. (B)177, p. 180.

[15]   *Ibid.*, p. (B)245, p. 226.

[16]   *Ibid.*, p. (B)259, p. 235.

[17]   *Ibid.*, p. (B)262, p. 236.

[18]   Albert Einstein, *The Meaning of Relativity* (Transl. by E. P. Adams), Methuen, London, 1922, p. 2.

[19]   Kant, p. (A)97, p. 130.

CHAPTER II

EUCLIDEAN AND RIEMANNIAN GEOMETRY

Euclidean geometry can be understood as a theory that derives from the construction of its fundamental concepts and whose theorems are true on the basis of its evident clarity [*Evidenz*]. The theorems of Euclidean geometry are therefore strictly valid for all objects of experience. On the other hand, the general theory of relativity shows that for the proportions of empirical objects it is not the Euclidean, but the Riemannian geometry which is valid.

The question therefore arises how these two apparently contradictory assertions can be reconciled. To answer this question, the basis for the evident clarity of Euclidean geometry as well as the empirical-physical arguments for the theory of relativity must be considered.

1. FORMULATION OF THE PROBLEM

Euclidean geometry is a theory that investigates the relations between ideal spatial figures. Its theorems can be derived, with the aid of logic, from a few fundamental propositions, the axioms. Since its discovery in Greek mathematics, Euclidean geometry has been regarded as a model theory whose axioms are true on the basis of its evident clarity and whose theorems are logically proven from these axioms. Thus, neither their objects nor their proof allows the theorems of Euclidean geometry to depend on experience. Nevertheless the theorems of geometry can be related to possible experience and applied to the dimensions of real objects: the relations asserted in geometry agree precisely with the empirical results obtained from the measurement of actual bodies.

Modern apriorism was the first to question the basis in experience for the clarity of the axioms and the validity of the theorems of geometry, and it was the first to provide an answer: the basic concepts of geometry can be defined only by ideal rules of construction, with the aid of which the fundamental figures can be constructed. But the rules of construction establish the general laws that are valid for geometrical concepts, i.e. the theorems of geometry. The rational clarity of the simplest geometrical principles, the axioms, derives from an intuition for the constructions on which the concepts are based. The procedures of construction, which make it at all possible to draw geometrical figures, thus establish the general laws which govern these figures. Therefore, lines, planes, triangles, etc. that are encountered in experience must necessarily obey the laws of Euclidean geometry.

The procedures of construction, and hence the fundamental concepts of geometry, at first appear to be arbitrarily chosen. In view of its application, a scientific geometry will be based only on principles of construction that are assumed, from prescientific experience, to be realizable at least in principle. We denote as 'pre-geometric' those properties of actual things that determine the possibility of constructing geometrical figures. The knowledge of 'pregeometric' properties requires no scientific geometry, and it can be taken by a qualitative, prescientific knowledge of experience.

The constructions of Euclidean geometry are possible in principle, if the objects of experience have the pregeometric property of 'unrestricted mobility'. Within the scope of this inquiry, 'unrestricted mobility' delineates the domain of empirical objects for which the theorems of Euclidean geometry are applicable. Since the things that one encounters in everyday life have this property of unrestricted mobility almost without exception, it is, in general, superfluous to state this property as a limitation upon the applicability of Euclidean geometry.

In the general theory of relativity the question of mobility of bodies is more rigorously discussed, within the framework of a physical theory. There the metric field investigated specifies the degrees of freedom that exist for the motion of a material body. Compared to knowledge of qualitative prescientific experience, the empirical circumstance is quantitatively and qualitatively more precisely and correctly comprehended in the theory of the metric field [*Führungsfeld*]. This theory has led to the result that material bodies do not, in general, have the property of unrestricted mobility, but rather a weaker property, which we shall call 'restricted mobility'. Thus, since the unrestricted mobility of material bodies, assumed for the applicability of Euclidean geometry, is not given, the theorems of Euclidean geometry are not applicable to the dimensions of empirical objects.

Under these circumstances there are two possibilities for empirical geometry: The first method of pursuing 'geometry', empirically, in spite of the circumstances described, consists in abandoning constructive definitions of the fundamental concepts. Instead one starts with certain empirically given objects and phenomena (e.g. rigid bodies or light rays), which one uses as standards [of length] or which one defines as straight lines. The relations between the positions and dimensions of material bodies obtained with these standards, are then interpreted as the theorems of an empirical geometry. This empiricist conception of geometry has seemed ever stranger since Gauss and Riemann, and finally led to the Riemannian geometry as formulated by Einstein in the general theory of relativity. The standards used are inertial paths of mass points, and light rays.

An alternative method of formulating empirical geometry is closely related to ideas developed in philosophical apriorism. In this case one takes Euclidean geometry as *a priori* and formulates the physical theory of the metric field with respect to this Euclidean metric, without regard to the question of empirical realizability. The metric field calculated in this way then no longer has the property of unrestricted mobility of bodies, but only that of restricted mobility. If one now regards this similarly qualitative property of restricted mobility as a 'pregeometric' property of the metric field, then one may ask what empirically realizable constructions of the fundamental concepts of geometry can still be stated. It can be shown that in a real metric field only a Riemannian geometry is possible. This Riemannian geometry agrees exactly with the empirical geometry of light rays and mass paths as used in the general theory of relativity.

Both of these possible methods of establishing an empirical geometry are based on two alternative formulations of the theory of the metric field. The first theory, stated by Einstein, relates all physical laws directly to the empirically founded Riemannian geometry, The theory of the metric field thereby becomes a theory that contains only observable quantities. It describes the metric field as it appears when measured with real standards. The second possible formulation of this theory relates the physical laws to a Euclidean representation space, which, however, cannot be measured with real standards. Only the metric field as calculated in the theory is observable, and it agrees exactly with the metric field calculated in Einstein's theory. This formulation of the theory of the metric field makes it possible then, by analogy, to carry the program for creating a foundation for geometry, as conceived in apriorism, over to the new circumstances.

2. THE FOUNDATION OF EUCLIDEAN GEOMETRY

2.1. *Greek Geometry*

Euclidean geometry is a system of theorems, which deal with the characteristics of certain fundamental spatial figures (points, lines, planes) and the relations between them (intersection, connection). The Greek mathematicians had discovered that the totality of these theorems can be logically derived from a small number of fundamental propositions, the axioms. The oldest surviving presentation of geometry as an axiomatic theory is the *Elements* of Euclid.

Euclid starts with a few definitions of geometrical figures, which, however, are not definitions in the strict sense but are to be taken as reminders for those already familiar with the meaning of the concepts. Thus, for example, Euclid defines "a straight line as a line that lies evenly with the points on itself". Inferences are not drawn from these definitions. The actual foundation of geometry consists of five axioms, which Euclid calls postulates. In these propositions the possibility of constructing certain fundamental figures is demanded. Since in Greek mathematics the solution to a geometrical problem consists in specifying a construction, the axioms of geometry must secure the possibility of fundamental constructions. In order to deduce the theorems of geometry from these axioms, several additional propositions are required. These, as for example the axioms of equality, are not, however, specifically part of geometry. Euclid demonstrated how the familiar theorems of geometry can be deduced from the postulates on a strictly logical basis. Until the 19th century, his representation was considered to be exemplary for every axiomatic theory.

The axioms of geometry cannot be proved and must precede the theory as unproved assertions. The question of the origins of these postulates already concerned Plato. It is related to the question of what it is that geometrical theorems deal with. Plato concludes that the lengths, angles, etc. treated by geometry are not the dimensions of actual objects, but that the objects of geometry are ideal geometrical figures, only mentally comprehensible, and that all proofs relate only to these ideal figures.[1] The actual lines, triangles and angles used in the proofs only serve for the purpose of demonstration.[2]

The truth of geometrical propositions thus derives not from experience but from insight into the essence of ideal figures. Depending on the degree of difficulty, this insight into a mathematical situation is either immediate, or it can be derived from demonstrations on actual things and from a dialectic argument. Plato defines this manner of cognition of an ideal situation as *anamnesis*, the recollection of knowledge that mankind enjoyed at an earlier time.[3]

Plato also questions to what extent pure geometry, recognized as true, is applicable to actual things. In order to understand the apparent correspondence between ideal-geometrical and actual-material forms, Plato takes recourse to his cosmology, in which the actual world was formed by its creator in the image of immutable notions. But since timeless ideas, to which the ideal geometrical forms belong, can be realized in the temporal only imperfectly, one must accept that an actual object assumes a geometrical property, for example that of being triangular, "with [only] some certainty".[4] The propositions of pure geometry and the relations among the measured dimensions of empirical bodies thus are only loosely connected. Because of this separation between ideal and actual objects, Plato has no problem of the applicability of geometry. For there is no doubt that geometry is empirically "valid with only some certainty". Plato never asserted that, beyond this, it is exactly valid in experience.

2.2. *Kant's Theory of Geometry*

The same phenomenon of geometry, which Plato attempted to interpret within the framework of the doctrine of anamnesis, also presented itself to Kant: a system of propositions that deal with ideal objects, points, lines and planes, and the relations between them. The truth of these propositions is based on their evidence, which either is directly present, in the axioms, or, for deduced theorems, is given by the respective proof. In the terminology of Kant, the axioms of geometry are synthetic *a priori* judgments. Synthetic, because of the content of these propositions does not follow directly from the definitions of the concepts used in these propositions; *a priori*, because the propositions of geometry are not valid on a basis of contingent experience, but rather are thought together with their necessity"[5], and so are valid apodictically.

Although the propositions of pure geometry, neither by subject nor by manner of foundation, have anything to do with experience, Kant considers them to be entirely and strictly applicable to objects of experience. In this Kant goes beyond Plato, who assumed that for the real world the propositions of geometry are valid only "with some certainty". By this is not meant that, for example, ideal triangles actually occur in reality, but rather that geometrical propositions are the more exactly valid, the more the empirical objects encountered approach ideal geometrical forms. Here Kant is in the tradition of modern science, which since Galilei and Kepler, starts with ideal exact laws, to which actual things conform exactly only in ideal limiting cases.

Kant asks how this synthetic *a priori* geometry, given him by tradition and of undoubted reality, is possible. "We have therefore some, at least uncontested, synthetic knowledge *a priori* and do not have to ask whether such knowledge is possible (for it is real), but only how it is possible."[6] If, with Kant, one understands by geometry, the "science which determines the properties of space synthetically, and yet *a priori*"[7], then the discussion of the question, how such a science is possible, necessarily leads to the question of the nature of space: "What, then, must be the representation of space, in order that such knowledge of it may be possible".[8]

In Kant's answer to this question, space is the form of all external phenomena and as such belongs to the conditions which make knowledge of objects at all possible.

If one requires that the synthetic propositions of geometry follow from the representation of space, then space must originally be an intuition, since no synthetic propositions can follow from a discursive concept, but only analytic ones. Furthermore, this intuition must be present in us prior to every perception of an object, i.e. it must be a pure, non-empirical intuition. For otherwise the propositions of geometry would not be apodictic, but only empirically valid.

Obviously this is possible only if the intuition of space as the formal nature of all objects is present in the perceiving subject. The manifold of perceptions becomes the phenomenon of things only in that the perceptions are brought into the form of space, i.e. ordered from the point of view of coexistence. For the subject, space is therefore an immanent form of perception by which the appearance of objects in space becomes possible, i.e. space belongs to the conditions of the possibility of experience of objects.

Space determined in this way has the property of 'transcendental ideality' as well as that of 'empirical reality'. By 'transcendental ideality' is meant that space itself is not an object, but belongs to the conditions which make the experience of objects possible. By 'empirical reality' is meant that all objects encountered in experience are in space, precisely because space is nothing other than the form of an external phenomena of objects of experience. This connection between space-as-pure-intuition and empirical space forms one of the presuppositions that makes it possible to relate pure and applied geometry.

"Geometry is the science which determines the properties of space synthetically, and yet *a priori.*"[9] Since space is the form of all external phenomena, the propositions of geometry refer to this pure intuition of space. They do not follow logically as analytical judgments from the definitions of geometrical concepts, but "from pure intuitions, (though mediated by understanding)".[10] These comments refer to arbitrary geometrical propositions, all of which are synthetic *a priori* judgments. But among these propositions there are some that are especially characterized by their evidence, which Kant calls axioms, in agreement with common usage. "These, insofar as they are immediately certain, are synthetic *a priori* principles."[11] They follow immediately from pure intuition. The remaining propositions of geometry can be derived from these axioms with the aid of logic and Kant similarly designates them as synthetic judgments, "for one can of course have insight into a synthetic proposition according to the principle of contradiction[12], but only by presupposing another synthetic proposition from which it can be made to follow, never in and by itself".[13]

An exact answer to the question of how the geometrical axioms can be characterized as immediately certain is thus required for the foundation of geometry. The axioms are concerned with the fundamental concepts of geometry and the relations between them. In geometry one encounters a kind of conceptualization that differs from the forming of an empirical concept or a philosophical concept given *a priori.* Whereas empirical concepts are obtained by abstraction from experience and philosophical concepts by explication or by definition from other concepts, geometric concepts are obtained by construction in pure intuition. For Kant to construct a concept means "to exhibit *a priori* the intuition which corresponds to the concept".[14] For the construction of a concept, a non-empirical, pure intuition is necessary. Thus one arrives at the concept of a triangle by constructing the intuition corresponding to it, either by mere imagination in pure intuition, or by drawing the figure on paper in empirical intuition.

The single figure which we draw is empirical, and yet it serves to express the concept, without impairing its universality. For in this empirical intuition we consider only the act whereby we construct the concept, and abstract from the many determinations (for instance, the magnitude of the sides and of the angles), which are quite indifferent, as not altering the concept 'triangle'.[15]

The fact that geometrical concepts are obtained by construction has the following basis: geometrical concepts refer to figures in intuition space, by which these concepts are represented in pure intuition. In a metaphysical exposition of this concept Kant says that space "is represented as an infinite given magnitude"[16], which is indeterminate with respect to the aggregate of its parts and hence constant within itself. Space is a continuous quantity. Spatial content (lines, surfaces, etc.) arise from prior decomposition and subsequent synthesis of quantum space. "I cannot represent a line to myself, however small, without drawing it in thought, that is, generating from a point all its parts one after another".[17] Geometrical figures are generated therefore by "the successive synthesis of the productive imagination".[18] They are magnitudes for which the representation of the parts precedes the representation of the whole, which can therefore be comprehended quantitatively. It is this quantitative character of geometric figures that makes possible the construction of the corresponding concepts. "For it is only the concept of quantities that allows of being constructed, that is, presented *a priori* in intuition."[19] Geometrical concepts relate to connected partial aggregates of quantitative space and can therefore be represented *a priori* in intuition, i.e. can be constructed.

The procedures, with the aid of which one constructs the intuition corresponding to the concepts, thus in a sense define the geometrical concepts. From the construction procedure for a triangle it should be possible, following Kant, to see immediately the theorem on the sum of angles. Kant expressly emphasized, however, that this theorem cannot be derived in the usual sense from the definition of a triangle alone, and it is not *a priori* an analytical judgment.[20] The axioms of geometry, and so the geometrical propositions that can "be immediately certain"[21], thus derive immediately from the construction procedures for the concepts, which the axioms treat. The remaining geometrical propositions can then be derived logically from these axioms.

These comments clarify in principle in which sense geometrical propositions are to be regarded as synthetic *a priori* judgments. The evidence for the axioms can be explained as an immediate insight into the principles of construction of the fundamental concepts. With that the synthetic as well as the *a priori* nature of the axioms is demonstrated. Kant, however, did not give explicitly the construction procedure for any geometrical concept. As a consequence, he was not able to clarify the evidence for any of the geometrical axioms. Presumably such proof was of no interest to Kant because he had no reason to doubt the evidence for the axioms of Euclidean geometry.

It is apparent even from this general discussion of the foundations of geometry that the *a priori* validity of Euclidean geometry is not a specific property of 'space' as a form of intuition, but rather derives from the constructions on which the geometrical concepts are based. Intuition space is merely a three-dimensional fixed manifold – a three-dimensional topological space[22] – in which the constructions necessary for the establishment of a geometry are undertaken. These constructions, and not the intuition space *per se*, are then responsible for the validity of geometrical propositions. Later we will consider the difficult question, which topological relations the intuition space may have.

The most important problem for our question is whether and to what extent this pure geometry, constructed in intuition space, is valid for objects of experience. Kant maintains that the propositions of Euclidean geometry are strictly valid in experience, and justifies this assertion. The justification takes place within the framework of establishing the nature of objects of experience. The nature of things is determined by the conditions of the possibility of experience, for

the conditions of the possibility of experience in general are likewise conditions of the possibility of the objects of experience, and for this reason they have objective validity in a synthetic *a priori* judgment.[23]

These synthetic *a priori* judgments characterize the objects of experience, because objects are always things already formed, in which the manifold of perceptions has been formed and interpreted from a certain point of view.

Thus, what we know of the principle of forming is also valid for that which is formed.

In the four principles of pure understanding, Kant characterizes the nature of things. The first of these is the 'principle of the axioms of intuition'. Along with other claims, it asserts the applicability of geometry to experience. The axioms of intuition are themselves the axioms of geometry. They are not posed for discussion here, only their applicability is. The principle of the axioms of intuition reads: "All appearances are, in their intuition, extensive magnitudes."[24] "All intuitions are extensive magnitudes."[25]

The proof, which Kant gives for this principle, at the same time illustrates its content; it starts with the conditions for the possibility of the experience of objects. The form of all external phenomena is the intuition of space. Phenomena reach consciousness by the same synthesis of manifolds whereby the representation of a certain space is produced. The representation of a certain space interval arises from the given quantity space by prior decomposition and subsequent synthesis. A spatial configuration is thus a magnitude for which the representation of the parts precedes the representation of the whole. Kant terms this an extensive magnitude. With regard to their spatial extension, all phenomena are therefore extensive magnitudes; for they came about as intuitions by the same synthesis by which a space interval is determined.

The significance of this proposition for the problem of the applicability of geometry consists in the following consideration: The figures of pure geometry arise in intuition space by a "successive synthesis of the productive power of imagination", that is, by consciousness of the synthetic unity of the spatial manifold. But space is the form of all phenomena and, as such, necessarily has empirical reality. With regard to their spatial extension the objects of experience similarly come about by a synthetic unity of the manifold of perceptions. Therefore, they are created in the same way as are geometrical figures. In pure geometry that synthesis, which collects the manifold of perceptions into spatial phenomena and thereby makes empirical objects possible, is executed in the pure intuition of space. In this way the figures of pure geometry come about.[26]

In principle, pure geometry can therefore be applied to the objects of experience. To do this one must merely execute in the empirical space the constructions on which the figures of pure geometry are based. That this is at all possible is demonstrated in the principle of the axioms of intuition. When, in this way, one has transposed the fundamental figures of Euclidean geometry (straight lines, planes, triangles, etc.) into the empirical space, then the propositions of Euclidean geometry, obtained in intuition space, are valid with full precision for the empirical magnitudes thus constructed. In this sense, we understand Kant's comments that "pure mathematics in its full precision is applicable to objects of experience"[27], and that the geometer can "be secured about the undoubted objective reality of his propositions".[28] Now we understand the meaning of the comment previously made, that geometry is only valid in an idealized limiting case. Empirical objects will satisfy the ideal requirements of planarity, of straightness of line, etc only to a certain extent. Geometric propositions about planes and straight lines will become more exactly valid, the more the constructions in empirical space approach the ideal requirements.

Space as the form of phenomena belongs to the conditions of the possibility of experience. Only in space is it possible to arrange the manifold of perceptions from the point of view of coexistence and to unify them as spatial phenomena. Space, the form of intuition necessary for experience, by which, in fact, it is constituted, is three-dimensional and topological. In addition, it is demonstrated in the principle of the axioms of intuition that it can be metricized, since by the very synthesis by which things enter consciousness these things can be quantitatively apprehended in their spatial extension. Space as form of all phenomena, constitutive for experience, is therefore a three-dimensional topological and metricizable space.

On the other hand, from these considerations, nothing precise can be said about the relations of space as a whole. One may assume that this structure of intuition space accommodates the given realities of the manifold of perceptions. If these allow themselves to be arranged in a simply connected space, then one will use it to order the perceptions; otherwise more complicated structures are considered. Since all perceptions immediately accessible to man can be arranged without difficulty in a simply connected space, we always draw on this simplest structure first. This problem first becomes acute in the discussion of cosmological questions.

Although three-dimensionality, metricability, and the relations of space as a whole are constitutive of experience, this is not the case for Euclidicity. The latter is, in fact, generated by the construction of the concepts relevant to Euclidean geometry. That Euclidean geometry, constructed in intuition space to begin with, is applicable to experience, is occasioned by the fact that the same geometrical constructions can be executed in intuition as well as in empirical space. This is demonstrated in the principle of the axioms of intuition. This principle not only ensures the validity of Euclidean geometry in experience, but also the empirical validity of geometry in general. The details of the propositions of geometry are immaterial for the proof of this principle; they depend on the principles of construction of geometrical concepts with which the propositions deal. Experience is therefore possible before one sets forth a scientific geometry that structures this experience in a particular way. Euclidean geometry is possible but not a necessary structuring of pre-geometric experience.

Since Kant only knew Euclidean geometry and its relevant concepts, he always chose Euclidean geometry as an example, demonstrating that it was valid not only in pure intuition, but also in empirical space. He mentions, for example, the proposition "that the straight line between two points is the shortest"[30], that "two straight lines cannot enclose a space, and with them alone no figure is possible"[31], and that the sum of angles in a triangle is equal to two right angles[32], from which it evidently follows that Euclidean geometry is meant.[33]

One must note, of course, that these propositions are not intended to imply that non-Euclidean geometries are not valid for experience. These were not even considered as alternatives, first of all because they were not known to Kant. The essential reason, however, is that for Kant geometry is a science that proceeds from the construction of its concepts, which in Kant's opinion is the case in Euclidean geometry. Whether it is possible to construct other concepts from which different axioms can be derived is not discussed. A pronouncement concerning the applicability of non-Euclidean geometries to experience can be made only after it has been clarified whether they are geometries in the sense advanced.

By his considerations, Kant showed how a scientific geometry is possible, which consists of synthetic *a priori* propositions, and which is fully applicable to experience. He assumed it to be self-evident that Euclidean geometry is a theory of this kind. Hence, Kant did not give a proof for this. It would consist in the explicit specification of rules of construction for the geometrical figures, on the basis of which the axioms of Euclidean geometry are immediately evident. The assertion that this geometry is fully applicable to experience is also only proved in principle. It is not shown how the constructions executed in intuition space are to be transposed to empirical objects, nor what the standards of empirical geometry are to be.

Subsequently both questions aroused great interest. The discovery of the logical possibility of non-Euclidean geometries raised the question whether the Euclidean axioms really were the evident and obvious propositions that Kant asserted them to be. The impossibility of realizing Euclidean geometry empirically shown in the theory of gravitation, has cast doubt on the strict empirical validity of Euclidean geometry, as asserted by Kant.

2.3. *Empiricism: Gauss, Riemann, Helmholtz*

Whereas in the Kantian sense geometry is a science that is based on the construction of its fundamental figures in intuition space as well as empirical space, the investigations of Gauss pointed to an entirely different direction. Gauss, as did Lobachevski and Bolyai at a later date, came to the conclusion that the fifth Euclidean axiom – the axiom of parallels – was logically independent of the remaining axioms. One can therefore replace it by an alternative assertion that contradicts this axiom, and thereby arrive at an axiomatic theory that is also without contradiction. Theories constructed in this manner are called 'non-Euclidean geometries'. In what sense this can still be considered to be geometry, however, is not immediately clear.

The investigations of Gauss on the inner geometry of curved surfaces and the constructive investigations of Riemann pointed to a solution of this question. Curved surfaces that are characterized by a constant measure of curvature have a particularly simple form of differential geometry on the surface. For the cases of positive and negative curvature, one obtains precisely both non-Euclidean geometries as the inner geometries of these surfaces.

The generalization of these considerations to three-dimensions suggests that, as for curved surfaces, a non-Euclidean geometry can be realized for three dimensional space, provided analogous circumstances prevail there. The generalization of differential geometry to arbitrary dimensions, introduced by Riemann, then led to the result that in spaces in which the curvature K is everywhere and in all directions constant – in the Riemannian spherical spaces – a non-Euclidean geometry is valid.

With this background it is understandable that Gauss as well as Riemann regarded it as a meaningful question to examine which of the three possible geometries is valid in empirical space. Here, the essential decision is made that it is meaningful to speak of an empirical metric structure of space, even without first engaging in geometry. In the conception of apriorism the geometry of space develops through the geometrical constructions executed in it.

For Gauss the systems of axioms of the Euclidean and non-Euclidean geometries are therefore equivalent mathematical structures, the validity of which for real space can only be empirically decided. In particular Gauss disputed the possibility that one of these geometries can be distinguished from the others by its inner evidence. In order to determine the geometry of real space, Gauss himself carried out an experiment by which the metric structure of empirical space was to be ascertained. One can decide between the three possible geometries by determining the sum of the angles in a triangle. Gauss therefore attempted to use conventional triangulation to measure with light rays the sum of angles in a large triangle formed by three mountain peaks.[34] No deviation from Euclidean geometry could be detected within the margin of error.

To what extent the result of this experiment is indeed a statement about space and to what extent it depends on the physical properties of light rays was not discussed by Gauss. He apparently assumed that light rays travel on the geodesics of a Riemannian spherical space, since otherwise the triangulation measurements would have no meaning. The difficult question, whether light rays are at all suited for the measurement of space and what the nature of measuring instruments must be, was not raised by Gauss nor, at a later date, by Riemann.

At the basis of this problem is the question, in which substantial sense Euclidean geometry and the two non-Euclidean geometries can be denoted as 'geometries', that is, what is actually meant by the obviously generalized concept of geometry. To be sure, these three geometries can be understood mathematically as a special case of Riemannian differential geometry; but the substantial meaning of Riemannian geometry will itself remain unclear until it is clarified why the fundamental concept of this theory, the line element ds, has the form d$s = \sqrt{g_{\mu\nu}\mathrm{d}x^{\mu}\mathrm{d}x^{\nu}}$, assumed hypothetically by Riemann.

These questions were decisively clarified by the investigations that H. v. Helmholtz devoted to the problem of space.[35] As Gauss and Riemann did, Helmholtz starts with the assumption that empirical space may have a given metric structure, and inquires what possibilities we have for empirical determination of this structure. Helmholtz takes the distance between two points as the fundamental concept of geometry. Empirically, distances can be measured only by comparison with a standard unit of length fixed somewhere in space. But for this purpose it must be possible to transfer the standard of length to the object investigated, that is, there must exist some bodies which can be used as measuring device, and which can be freely moved in space. However a body can be used as a measure only if it is rigid, i.e. if the relative distances between different points, which are marked on the body, are not changed by moving the body. Thus, to be able to measure distances one must be able to establish the congruence of lines; this is only possible if freely movable bodies exist, which can be used as transportable standards. The existence of freely movable rigid bodies is one of the necessary conditions that make an empirical geometry possible at all. Under these circumstances one may ask whether these conditions for the possibility of empirical geometry do not, from the first, already establish certain structures of the space that one is empirically measuring. Helmholtz demonstrated that this is in fact the case, and he explicitly specified the relevant structures of space.

These considerations derive from the following mathematical basis (Principle of Helmholtz): We consider, instead of a rigid body, a system of points in a bounded region. The free mobility of a body then corresponds to an isometric mapping of the bounded region on itself. First of all we shall assume, however, that the spatial region is sufficiently small, so that only arbitrarily small bodies are freely movable. The principle of Helmholtz then asserts[36]: If sufficiently small spatial regions can be mapped on themselves isometrically by a point rotation, then the space is Riemannian. The line element then has the form $ds^2 = g_{\mu\nu}\, dx^\mu dx^\nu$. If in addition it is assumed that point systems of finite extension can be mapped on themselves by point rotation, then the space investigated is a Riemannian spherical space. The curvature of space, K, is then a constant, independent of direction and position. The geometry of such a space is then either Euclidean ($K = 0$) or one of the two non-Euclidean geometries ($K \lessgtr 0$).

These mathematical relations indicate a possibility of comprehending Euclidean and non-Euclidean geometries from a substantial point of view as generalized 'geometries'. For if the finite point systems are freely movable, then the congruence of finite distance is defined in this space. Conversely, if one starts with the concept of distance and the necessarily associated free mobility of finite point systems, then one is led to absolute geometry; the question of the curvature of space, K, remains open. Euclidean geometry and the non-Euclidean geometries can therefore be understood, with regard to content, as distance geometries.[37] For a general Riemannian space, in which only free rotation of arbitrarily small space elements occur, only the congruence of infinitesimal distances can be defined in a corresponding manner. The Riemannian generalization of geometry thus consists in the assumption that distances are only defined for infinitesimal contiguous points.

From Helmholtz's principle of homogeneity one obtains the following result for empirical geometry: If sufficiently small, rigid bodies, freely movable in space, are empirically given, then the geometry of the space which one measures with these rigid bodies as standards is the geometry of a Riemannian space. If, in addition, rigid bodies of finite extension are freely movable, then the geometry measured with these bodies is the geometry of a Riemannian spherical space. The value of the curvature of space, K, cannot be predicted from assumptions made about the measuring standards.

Helmholtz started with the empirically nearly valid observation that freely movable rigid bodies of finite extension exist. If we use these bodies as measuring standards for the determination of congruences, then we are necessarily led to absolute geometry. The question of the precise value of the curvature of space K can then only be resolved empirically, for example by determining the sum of angles for a triangle. In this manner the problem of empirically determining the geometry of real space can be divided into two steps: the first is to establish empirically that there exist freely movable rigid bodies of finite extension. Using these bodies as measuring standards necessarily leads to a Riemannian geometry of constant curvature K. The value of K must then be experimentally determined in the second step.

The significance of these statements for the geometry of empirical space is still unresolved as long as it is not clarified how one tests bodies for rigidity. This can be accomplished only by marking, on the body investigated, at least four points that do not lie in a plane, and by comparing the distances between these points before and after a certain motion. If the distances remained unchanged, and thus if the system of points has been transformed to another congruent system of points, then one can designate the body as rigid. But to demonstrate the congruence of two point systems one must have freely movable measuring standards the rigidity of which, in turn, could only be demonstrated by congruence measurements. There apparently exists no method for empirically testing the rigidity of a measuring standard.

Under these circumstances one can conversely assert of an arbitrary, freely movable body that it is rigid. One can then use this body, *defined as rigid*, as measuring standard for the mensuration of space. If the body is sufficiently small, then the measured geometry of space is Riemannian; if furthermore it has finite extension, then one obtains the geometry of a Riemannian spherical space. The numerical value of the curvature K depends on the measuring body which has specifically been defined as rigid or, more precisely, on the interaction which this measuring body undergoes with any possible force fields in space. The geometrical propositions which one obtains from the mensuration of space therefore do not deal with the properties of empty space, but rather with the measuring standards, defined as rigid, and their reciprocal interaction with various force fields.[38]

In his definition of rigid bodies, Helmholtz assumes that a structure that is known to be physically undeformable can be designated as rigid. The undeformability must therefore be established by physical measurement. But as long as one does not exactly know which force fields possibly act on the 'rigid' body, one cannot appraise what influence this dynamic approach, used for the definition of rigidity, has on the geometry of space. The dynamic-physical definition of rigid bodies therefore only determines a specific measuring standard from particular physical situations. As we have seen, one evidently could designate an arbitrary body as rigid, and thereby obtain, for example, any specific desired geometry; this, however, would entail a change in the physical laws.[39]

The question of the geometrical structure of empty space can, therefore, not be answered empirically. That geometry that one finds experimentally depends on the definition of the measuring standard and on the changes that the measuring standard experiences under force fields present in space. The empiricism of the geometry of empty space obviously has no reasonable meaning.

2.4. *Conventionalism and Apriorism: Poincaré, Dingler, Lorenzen*

For the question of the foundations of geometry, the discovery of non-

Euclidean geometries raised the problem of showing one of the possible geometries to be the true geometry. The investigations of Helmholtz have shown that the substantial point of view of a distance geometry can only lead to absolute geometry, and leaves the decision among the three possible geometries unresolved. In general there are two viewpoints by which one can attempt to make this decision:

(1) One of the three geometries is distinguished from the others by its own evidence; its propositions are interpretable as synthetic judgments *a priori* in the sense of Kant. In the axiomatic formulations of the geometries, no such preferred status can be recognized for any of the possible geometries. Furthermore, since Kant has only very incompletely treated the question of the foundation of the apriority of geometrical propositions, this possibility has generally been rejected since Gauss.

(2) Similarly it proved impracticable to recognize any one of the geometries as actualized in real space. All empirical investigations deal with measuring standards and force fields, but never with space itself, or with the geometry valid in it. Therefore an empirical decision between the Euclidean and non-Euclidean geometries is not possible.

Against this background Henri Poincaré[40] concluded that geometry is not a science of experience, nor does it consist of self-evident propositions. Geometry (characterized by a particular system of axioms) is a theory which, to begin with, defines the objects (points, lines, planes) and relations (intersection, connection) with which it deals. The only requirement that can be demanded of such a system of axioms — which implicitly expresses definitions — is its freedom from contradiction, its self-consistency. This point of view was to a large extent adopted by Hilbert in his *Grundlagen der Geometrie*.[41]

The question of a true geometry is therefore meaningless for Poincaré. The geometrical axioms are definitions of concepts and rest on convention. From the totality of possible geometries, one will, however, choose one that is distinguished by its mathematical simplicity and its expediency for applied geometry. Poincaré was convinced that these two aspects unambiguously characterize Euclidean geometry. Simplicity is obviously present when Euclidean geometry is compared with the non-Euclidean. With regard to expediency, Poincaré points out that bodies, physically recognized as rigid, when used as measuring standards, lead to Euclidean geometry with adequate exactness. The important and difficult question whether this can be attributed to chance, or whether there is some basis why physically rigid bodies permit a realization of Euclidean geometry, is not posed by Poincaré. It was not until the general theory of relativity that the question could be considered.

Fundamentally the empirical realization of one of the possible geometries does not pose a new problem. For if one regards geometrical axioms as implicit definitions of their concepts, then this point of view can be carried over to applied geometry. One describes an empirical line as a Euclidean straight line precisely when it obeys the relevant propositions of Euclidean geometry, and correspondingly a surface as a Euclidean plane when it obeys the propositions for planes. In this manner the realization of a particular system of geometrical axioms in experience, is always possible – provided, of course, that one encounters surfaces and lines which, at least approximately, can be designated as planes and straight lines.[42] In this way one avoids the arbitrariness of determining that a particular empirical object is a straight line (for example, the path of a light ray).

Similarly the arbitrariness of setting up an empirical standard of measurement can be avoided if one uses an analogous procedure for geometrically defining rigid bodies, which can then be used as standards for measurement of space.

Already with Helmholtz, we find the comment that in the absence of physics, the rigidity of a body can only be decided with the aid of geometrical propositions.[43] But if one defines by the propositions of Euclidean geometry whether a body is to be regarded as rigid, then a body rigid in this sense, when used as a standard, will always lead to the propositions of Euclidean geometry.[44]

This program can be carried out if one starts, for example, with Hilbert's axiomization of Euclidean geometry, which is based on the fundamental concepts of point, straight line, plane, intersection, connection, orthogonality and parallelism and which can be used to define the concept of congruence.[45] The parallel congruence of two pairs of points A, B and A', B' can be defined by the existence of a parallelogram with the corners A, B, A', and B'; the reflection congruence of two pairs of points A, B and A', B' can be defined by a parallelogram whose diagonals are orthogonal. In general, two point pairs A, B, A', B', are termed congruent if there exists a third pair X, Y, that is reflection- and parallel-congruent with each of the first two pairs.

With a congruence concept defined in this way, it is possible geometrically to define the rigidity of a body. A body is designated as geometrically rigid when every pair of points X, Y marked on it transforms for every motion of the body into a congruent pair of points X', Y'. Applied to a number of bodies, one then obtains a criterion of rigidity. If one uses such geometrically rigid bodies as measuring standards for measurement of space, then one necessarily verifies the propositions of Euclidean geometry. On the basis of the congruence concept employed and the geometrical definition of rigidity, this is logically apparent.

Thus for the problem of the foundation of geometry, this step appears to have contributed little. With the aid of the system of axioms – in this case for Euclidean geometry – measuring standards are defined with which the geometry derived from the axioms must necessarily be verified empirically. A decision in favour of one of the possible geometries is obviously not possible in this manner; the stated procedure merely appears to confirm the suspicion that the choice of geometry is entirely arbitrary, and that even the standards can be defined so that the arbitrariness is transferred to the empirical geometry. But by comparison with the previously mentioned empirical procedure for establishing a measuring standard and the quality of straightness, a certain methodological advantage is attached to such a procedure. In establishing the rigid measuring standard or the straight line, we do not depend on contingent physical objects and phenomena (physically rigid bodies, light rays) but rather we define with geometrical concepts when a body is to be designated as (geometrically) rigid, and when a line is to be called straight. The question, which geometry results empirically, then no longer depends on fortuitous physical properties of the standards used, but can be precisely predicted on the basis of the definitions of the measuring standards or straight lines.

By such a procedure, geometry does not, however, become a system of synthetic *a priori* judgments in the sense of Kant. Already Helmholtz noticed that in this case merely the measuring standards are defined, with the aid of geometrical propositions, such that the geometry measured has the desired structure. But then the axioms of geometry are not propositions the validity of which is based on their self-evidence. In the Kantian sense, this is only possible if the fundamental concepts of the theory are constructed in pure intuition, and if the validity of the axioms can be immediately comprehended on the grounds of the rules of construction used. The systems of axioms here discussed, on the other hand, contain undefined concepts, which are only implicitly defined by axioms with respect to other similarly undefined relations. Euclidean geometry, furthermore, is in no way distinguished from other systems of axioms.

The supposition that the Kantian program may be possible after all for the foundation of geometry was first advanced by Hugo Dingler.[47] Dingler also explicitly specified some rules of construction for geometrical concepts. More recently Paul Lorenzen elaborated this approach and brought Dingler's rules of construction into a mathematical form by which the connection with the axiomatic formulation of geometry can be recognized. Dingler and Lorenzen have maintained that the direction taken by them unambiguously leads to Euclidean geometry.

The essential idea, which is added to the considerations so far presented, is that the geometrical fundamental concepts of plane, straight line, point, etc., are not only formally established by axioms, but also precisely, and concretely meaningful. Such substantial concepts presumably also form the basis of the definition of geometrical figures given by Euclid. For example, a straight line is defined as a line which lies evenly with the points on itself. Dingler attempts to define the concept of the plane by specifying rules by which a plane can be practically produced with any desired precision. This can be done in the following manner: Two bodies A and B are ground on each other; in this way a convex and a concave spherical surface result. If one now takes a third body, C, which one grinds on A as well as B, then plane surfaces result on the bodies A, B, and C.

This procedure is not intended to be concretely executed, but rather the concept of an ideal plane or the notion of a plane is to be characterized by these rules of construction. That one can never produce an ideal plane by the mutual grinding of real bodies is immaterial. It is only intended to give a rule of construction for a surface whose result, ideal and technically never achievable, is to characterize the plane. Dingler terms this method of establishing and clarifying a concept an operative definition.

Lorenzen has pointed out that in this and other operative definitions, which Dingler sets forth, principles of homogeneity are contained as essential factor. The grinding procedure, for example, indicates nothing other than that all points of the plane are indistinguishable. The statement of homogeneity contained in it can be defined as follows: If E is a plane, P

and P', are points, and $A(E, P)$ is a geometrical proposition that deals with E and P, then it is required that for every P and P'

$$P \in E \wedge P' \in E \wedge A\,(E, P) \to A\,(E, P')$$

is valid. Thus, if $A(E, P)$ is valid for a point P in the surface E, then the same proposition A shall also be valid for every other point P' which is in E.

By such principles of homogeneity, it is possible to define the concepts of plane, straight line, parallelism, orthogonality. If one further adds an explanation for the topological relations of incidence and of the notion of between, then all fundamental concepts and relations of geometry are defined. From these definitions and from the principles of homogeneity follow at once all the axioms of Euclidean geometry.

The empirical realization of Euclidean geometry then poses no new problem. One merely has to explain with operatively-defined concepts the concept of congruence whereby geometrically-rigid bodies then become definable. The use of such bodies as measuring standards then necessarily leads to a Euclidean empirical geometry. With the aid of geometrically-rigid bodies one can examine surfaces for their planarity. This can be carried out more simply by directly using Dingler's grinding procedure. It is important to note that it is now possible, with the aid of the Dingler-Lorenzen definition, to decide whether an individual surface is planar — without having to assume the similarity of a sufficiently large number of surfaces.

If one understands, with Dingler and Lorenzen, operative definitions of geometrical concepts to be a realization of the Kantian requirement that the fundamental concepts of geometry be constructed in pure intuition, then one interprets the axioms of Euclidean geometry as synthetic judgments *a priori*. Their truth derives from the rules of construction that produce a geometrical figure, or rather that operatively define ideal form. The carryover of this pure geometry into reality then takes place by the definition of the geometrically-rigid body, which is used as a measuring standard of empirical geometry. The axioms, first only valid in pure geometry, are then necessarily also valid for all objects of experience measured with this measuring standard. Since the geometrical concepts of plane, straight line, etc. are defined by rules of construction, one is compelled to carry out the realization of these forms in experience precisely in accordance with these rules. Kant had already proposed in the principle of the axiom of intuition that the spatial can be measured, and that thereby empirical geometry is possible. Only, the question of how the construction of the concepts is to be concretely executed in intuition was never considered by Kant. We may therefore suppose that Kant's deductions are supplemented by the considerations of Dingler and Lorenzen.

The foundations of geometry, established by this method, do not pick out a particular geometry. Rather, every theory can now be designated as geometry, provided its fundamental concepts can be obtained in pure spatial intuition, and provided its axioms necessarily follow from these rules of construction. The transfer of such a theory into empirical space is then always possible. The preferential status of Euclidean geometry existed for Kant presumably because for this theory he could imagine a constructive definition of its fundamental concepts, and because he saw no alternative. Similarly, the Dingler — Lorenzen foundation of Euclidean geometry by the principle of homogeneity allows no alternative to the construction of the Euclidean fundamental concepts.[48] From this one may not conclude, however, that only Euclidean geometry is based on self-evident axioms and that, for suitable transposition of the concepts to empirical space, the Euclidean geometry is necessarily valid for all objects of experience. Rather, this will be the case for every theory of spatial figures whose fundamental concepts can be operatively defined.

3. THE THEORY OF THE METRIC FIELD

3.1. *Einstein's General Theory of Relativity*

The motion of a mass point that is subjected only to gravitational forces is described in classical Newtonian mechanics by the potential equation for the gravitational potential

$$(1) \qquad \Delta\Phi = 4\pi\, k\rho$$

and by the equation of motion for a mass point

$$(2) \qquad \frac{d^2 \mathbf{r}}{dt^2} = -\,\text{grad }\Phi$$

where $\rho(x, y, z)$ is the density of the masses that produce the gravitational field ϕ, and k is the gravitational constant. The function $\mathbf{r} = \mathbf{r}(t)$ calculated from Equations (1) and (2) then gives the orbit of the mass-point considered. For the complete determination of $\mathbf{r}(t)$, boundary values must be assumed in equation (1) for $\mathbf{r} \to \infty$, and in equation (2) the initial values $\mathbf{r}(t_0) = \mathbf{r}_0$ and $\mathbf{r}'(t_0) = \mathbf{r}'_0$ must be assumed.

This theory is empirically completely verifiable. The Euclidean coordinates x, y, z, of a mass point can be experimentally determined for every point in time, t. The reason for this is that, within the framework of this theory, light is not affected by gravitation since it has a negligible rest mass. Light rays can therefore be described by straight lines and can at any time be used for the measurement of the empirical Euclidean space, in which the orbits of the mass points then assume the form $\mathbf{r} = \mathbf{r}(t)$. The complete empirical verification of this theory, however, is only possible on the hypothesis that light propagates instantaneously and can thus be used for the empirical realization of an absolute time, t.

In spite of its inner consistency, the Newtonian theory must be supplemented in several respects. The finiteness of the speed of light compels one to generalize this theory, in the sense of the special theory of relativity, for speeds that are comparable to the speed of light. Furthermore, it has been demonstrated empirically that light rays are affected by gravitational fields, as revealed by the deflection of light rays passing close to the sun. This has two important consequences for the theory of gravitation. Under very general assumptions the mathematical conclusion follows that the gravitational field cannot be a scalar field ϕ, but must be a tensor field $g_{\mu\nu}$. Furthermore, the deflection of a light beam, by a gravitational field has the consequence that the possibility of empirically realizing a Euclidean representational space in the theory is no longer at our disposal, since everything in the world (matter and radiation) is affected by gravitational forces. If therefore we require of a theory that all its stages of definition have experimentally verifiable meanings, then it is reasonable to refer the propositions of the theory not to the unobservable Euclidean coordinate axes, but to the actual paths of light rays and mass points, or to magnitudes which can be determined from these. In 1916 Einstein succeeded in carrying out this generalization of the Newtonian theory of gravitation. The result was the general theory of relativity.

In this theory the motion of a mass point is described by Einstein's field equations for the gravitational tensor field $g_{\mu\nu}$

$$(3) \qquad R_{\mu\nu} - \tfrac{1}{2} R g_{\mu\nu} = - \kappa T_{\mu\nu}$$

and by the equation of motion for the path $x^{\mu} = x^{\mu}(\tau)$ of a mass point.

$$(4) \qquad \frac{\mathrm{d}}{\mathrm{d}\tau} \left(g_{\mu\nu} \frac{\mathrm{d}x^{\mu}}{\mathrm{d}\tau} \right) = \frac{1}{2} \frac{\partial g\alpha\beta}{\partial x^{\nu}} \frac{\mathrm{d}x^{\alpha}}{\mathrm{d}\tau} \frac{\mathrm{d}x^{\beta}}{\mathrm{d}\tau}$$

where κ is the relativistic gravitational constant $\kappa = 8\pi k/c^4$

$$(5) \qquad c^2 \, \mathrm{d}\tau^2 = g_{\mu\nu} \, \mathrm{d}x^{\mu} \, \mathrm{d}x^{\nu},$$

$T_{\mu\nu}$ is the energy-momentum tensor of matter and the radiation field, $R_{\mu\nu}$ and R are magnitudes that can be expressed by $g_{\mu\nu}$ and its derivatives thus:

$$R_{\mu\nu} = g^{\alpha\beta} \left\{ \frac{1}{2} \left(\frac{\partial^2 g_{\nu\alpha}}{\partial x^{\mu}\, \partial x^{\beta}} + \frac{\partial^2 g_{\mu\beta}}{\partial x^{\nu}\, \partial x^{\alpha}} - \frac{\partial^2 g_{\nu\beta}}{\partial x^{\mu}\partial x^{\alpha}} - \frac{\partial^2 g_{\mu\alpha}}{\partial x^{\nu}\, \partial x^{\beta}} \right) \right\}$$

$$- g_{\sigma\rho} \, \Gamma^{\rho}_{\mu\alpha} \Gamma^{\sigma}_{\nu\beta} + g_{\rho\sigma} \, \Gamma^{\rho}_{\mu\beta} \Gamma^{\sigma}_{\nu\alpha}$$

$$R = g^{\mu\nu} R_{\mu\nu}, \qquad \Gamma^{\nu}_{\alpha\beta} = \frac{1}{2} g^{\mu\nu} \left(\frac{\partial g_{\alpha\mu}}{\partial x^{\beta}} + \frac{\partial g_{\beta\mu}}{\partial x^{\alpha}} - \frac{\partial g_{\alpha\beta}}{\partial x^{\mu}} \right)$$

From Equation (4) it is apparent that the force-free paths of mass points are given by paths $x^{\mu}(\tau)$, which can be understood as geodesics in a Riemannian space whose metric tensor is given by $g_{\mu\nu}$. In principle, therefore, this metric tensor can, conversely, also be determined by the empirical paths of mass points and light rays. In detail, however, this requires a very exact analysis of the respective processes investigated.[49] Because of this dynamic definition, the metric tensor $g_{\mu\nu}$ is also designated as the metric field (*Führungsfeld*).[50] This metric field, which can always be measured with real clocks and measuring standards, is now conceived to be the metric tensor of a Riemannian geometry. All physical processes are formulated with reference to this Riemannian geometry, that is, the fundamental equations of physics are to be written as covariant equations in a Riemannian space.

This is especially true of the field equations of gravitation. Einstein's field equation (3) is therefore a covariant differential equation for the $g_{\mu\nu}$-field and relates it to the energy-momentum tensor $T_{\mu\nu}$. From the point of view of field Equation (3), $g_{\mu\nu}$ appears as the gravitational potential that results from a matter distribution with the energy-momentum tensor $T_{\mu\nu}$. The equivalence of the metric field, dynamically defined by equation (4), and the gravitational field $g_{\mu\nu}$ defined by (3), is a physical circumstance grasped by Einstein's theory. The interpretation of the $g_{\mu\nu}$-field as a metric tensor of a Riemannian geometry and the all-embracing requirement of covariance for all physical equations, on the other hand, is a consequence of the methodological requirement, endorsed by Einstein, that a physical theory is permitted to contain observable quantities only. This requirement is satisfied in the general theory of relativity. It describes physical processes in gravitational fields as they show themselves when observed with actually existing clocks and measuring standards, and thereby refers explicitly to the possibility of observation and the measurement process.

The theory has been empirically corroborated by very precise measurements on the behaviour of bodies and light rays in gravitational fields. For the case when no matter is present to generate a gravitational field, and thus when $T_{\mu\nu} = 0$, one obtains from Einstein's field Equation (3) and suitable boundary conditions the result $g_{\mu\nu} = \eta_{\mu\nu}$, where $\eta_{\mu\nu}$ is the metric tensor of a pseudo-Euclidean space. It is the metric used in the special theory of relativity. It can be realized empirically under the assumptions mentioned, since, from Equation (4), the inertial paths of mass points and light rays are straight lines when $g_{\mu\nu} = \eta_{\mu\nu}$. The general theory of relativity then completely goes over into the special theory of relativity.

The general theory of relativity is based on the same epistemological concept as the special theory of relativity. Physics describes phenomena of nature as they appear when measured with real measuring standards and clocks. It is therefore a theory that contains observables exclusively. In principle there is in addition the possibility of admitting unobservable magnitudes into the theory as well. One could, for example, retain a Euclidean mathematical representation space and formulate the equations of motion of bodies in a gravitational field within it. The Euclidean coordinates would then be unobservable parameters, and the theory would no longer be completely verifiable. The precise properties of unobservable magnitudes are not prescribed by empirical data, however, so that new aspects would need to be drawn upon for their determination.

3.2. *Theory of Gravitation in Flat Space*

The generalization of the Newtonian theory of gravitation in the sense of the special theory of relativity, together with the requirement that all well defined points occurring in the theory be empirically realizable, leads to Einstein's theory of gravitation. If one denies the methodological requirement of consistent experimental realizability, then one arrives at a Lorentz-invariant theory of the gravitational field. The potential equation for the scalar Newtonian potential

$$(1) \qquad \Delta \Phi = 4\pi k \rho$$

must then be replaced by a Lorentz-invariant field equation. The experimental observation of light deflection compels one to use, under very general conditions, a tensor field instead of a scalar field. Instead of the scalar mass density ρ, the energy-momentum tensor of matter $T_{\mu\nu}$ is adopted in a Lorentz-invariant theory. If one further notes that the energy of the free tensor field $\psi_{\mu\nu}$ should be positive definite, then the simplest generalization of the potential Equation (1) is:

$$(6) \qquad \Box^2 \psi_{\mu\nu} = f(T_{\mu\nu} - \tfrac{1}{2} \eta_{\mu\nu} T)$$

with the condition:

$$(6a) \qquad \psi_{\mu\nu}{}^{,\nu} = \tfrac{1}{2} \psi^{\lambda}_{\lambda,\mu}$$

where $T = T_{\alpha\beta} \eta^{\alpha\beta}$ and f is a coupling constant. Equation (6) can be derived from the Lagrangian

$$(7) \qquad L(x) = \tfrac{1}{2} \left[\psi_{\mu\nu,\lambda} \, \psi^{\mu\nu,\lambda} - 2\,\psi_{\mu\nu,\lambda} \, \psi^{\lambda\nu,\mu} + 2\,\psi_{\mu\nu}{}^{,\mu}\,\psi_{\sigma}{}^{\sigma,\nu} \right.$$

$$\left. - \psi^{\sigma}{}_{\sigma,\lambda}\,\psi^{\nu,\lambda}_{\nu} \right] + fT^{\mu\nu}\,\psi_{\mu\nu}$$

by variation with the field quantities $\psi_{\mu\nu}$, if one also takes (6a) into account. The first term of the Lagrangian (7) corresponds to the free $\psi_{\mu\nu}$ field, the second to the interaction with matter.

The relativistic generalization of Newton's equation of motion

$$\frac{d^2 \mathbf{r}}{dt^2} = -\,\mathrm{grad}\,\Phi$$

can similarly be obtained from a Lagrangian. If the Lagrangian for N free mass points,

$$L^M = -\sum_{\alpha=1}^{N} m_\alpha c^2 \int d\tau_0$$

where $c^2\,d\tau_0^2 = \eta_{\mu\nu}\,dx^\mu dx^\nu$, is added to the Lagrangian of the interaction of these mass points with the $\psi_{\mu\nu}$-field

$$L^W = \int d^4 x f T^{\mu\nu}\,\psi_{\mu\nu} = f\sum_{\alpha=1}^{N} m_\alpha \int d\tau_0 \,\frac{dx^\mu_\alpha}{d\tau_0}\frac{dx^\nu_\alpha}{d\tau_0}\,\psi_{\mu\nu}\,[x_\alpha(\tau_0)]$$

then one obtains the equation of motion of the mass point α when $L^M + L^W$ are varied with the world line $x^\mu_\alpha(\tau_0)$ of this body. The equation of motion is then[51]

$$(8) \qquad \frac{d}{d\tau_0}\left[(\eta_{\mu\nu} - 2f\,\psi_{\mu\nu})\frac{dx^\mu}{d\tau_0} \right] = -f\,\frac{dx^\alpha}{d\tau_0}\frac{dx^\beta}{d\tau_0}\,\psi_{\alpha\beta,\nu}$$

$$- \frac{d}{d\tau_0}\left[\eta_{\mu\nu}\frac{dx^\mu}{d\tau_0}\frac{f}{c^2}\,\psi_{\alpha\beta}\frac{dx^\alpha}{d\tau_0}\frac{dx^\beta}{d\tau_0} \right]$$

in which the mass m_α of the body no longer appears. This circumstance, characteristic for the gravitational field, is denoted as the equivalence principle.[52] It is based on the fact that the $\psi_{\mu\nu}$-field is directly coupled to the energy-momentum tensor of matter.

The connection between this Lorentz-invariant theory of gravitation and Einstein's covariant theory of gravitation becomes apparent in the following way: In Einstein's theory, all laws of nature are referred to the metric $g_{\mu\nu}$ that can be determined by the measurement of the motion of mass points in the gravitational field. The guiding metric field $g_{\mu\nu}$, is therefore defined by the equation of motion. This equation of motion (8) can be regarded as the differential equation for a geodesic $x^\mu(\tau)$ in a Riemannian space with the metric tensor:

$$g_{\mu\nu} = \eta_{\mu\nu} - 2f\,\psi_{\mu\nu}$$

If one replaces $\psi_{\mu\nu}$ by $g_{\mu\nu}$ in Equation (8), one obtains, to the terms of order f^2, the differential equation

$$(9) \qquad \frac{d}{d\tau}\left(\frac{dx^\mu}{d\tau}g_{\mu\nu}\right) = \frac{1}{2}\frac{\partial g_{\alpha\beta}}{\partial x^\nu}\frac{dx^\alpha}{d\tau}\frac{dx^\beta}{d\tau}$$

which agrees precisely with Einstein's equation of motion in a gravitational field $g_{\mu\nu}$. τ is again the proper time (*Eigenzeit*) in Riemannian space defined by Equation (5). On the basis of the equation of motion (8), the metric field that can be measured by the path of a mass point is therefore given in the linear approximation in f by $g_{\mu\nu} = \eta_{\mu\nu} - 2f\psi_{\mu\nu}$.

If in the field Equation (6), one similarly replaces $\psi_{\mu\nu}$ by $g_{\mu\nu}$ and defines $\kappa = f^2$, then the field equation for the gravitational potentials $g_{\mu\nu}$ is obtained

$$(10) \qquad \Box^2 g_{\mu\nu} = -2\kappa\left[T_{\mu\nu} - \tfrac{1}{2}T\eta_{\mu\nu}\right]$$

which agrees precisely with the linear approximation to Einstein's field equation

$$(3a) \qquad R_{\mu\nu} = -\kappa\left[T_{\mu\nu} - \tfrac{1}{2}Tg_{\mu\nu}\right],$$

which one obtains for weak gravitational fields, if, in addition, the permissible coordinate systems are restricted in suitable fashion. In this linear approximation, Einstein's gravitational theory is therefore exactly equivalent to the Lorentz-invariant theory of gravitation here presented. The physical circumstance, that the metric field, defined by the equation of motion (9), is equal to the gravitational potential calculated from field Equation (10), is taken into account as in Einstein's theory.

The generalization of the Lorentz-invariant theory of gravitation, so that the field equations appearing in the theory are equivalent to the non-linear field equations of Einstein, creates no difficulties in principle.[53] Starting with a considerably more complicated Lagrangian than that given in (7), one obtains by variation with the field the non-linear field equations of the gravitational field. At first these are formulated in a pseudo-Euclidean space with the metric tensor $\eta_{\mu\nu}$. The transition to the covariant equations of Einstein then occurs by eliminating $\eta_{\mu\nu}$ and by relating the field equations to the metric $g_{\mu\nu}$ that is defined as the guiding metric field by the equations of motion.

In a flat space-time continuum, the theory of gravitation can therefore be completely formulated as a Lorentz-invariant field theory of gravitation. The discussion of the equation of motion shows, however, that only the components of the metric field $g_{\mu\nu}$, but not the pseudo-Euclidean coordinates can be experimentally determined from the motion of one or more mass points. If one acknowledges Einstein's requirement that all points of definition occurring in a theory must be empirically realizable, then the pseudo-Euclidean metric $\eta_{\mu\nu}$ must be completely eliminated from the field equations, and must be expressed by the observable metric field $g_{\mu\nu}$. In this way one arrives at Einstein's theory of gravitation.[54]

The pseudo-Euclidean metric field cannot, in general, be detected. It therefore appears in the theoretical formulation of the theory of gravitation as a system of hidden parameters, which, on the basis of the very theory in the formulation of which they are used, can never be measured. But the impossibility of a complete empirical verification of theories that use hidden parameters precludes any check, by empirical criteria, on the great arbitrariness in the choice of such hidden parameters. The question is still open as to why specifically the metric $\eta_{\mu\nu}$ is to be used in the formulation of the theory of gravitation rather than some other, similarly unobserved, metric $\tilde{g}_{\mu\nu}$.

The reasons to prefer the $\eta_{\mu\nu}$-metric are therefore not purely physical. The Euclidean space will tend to enjoy a preferred status compared with the Riemannian space because Euclidean geometry preceded Riemannian geometry, both historically and systematically. The apriority of Euclidean geometry discussed above clearly allows this privileged position to be recognized.

On the other hand, the metric field $g_{\mu\nu} = \eta_{\mu\nu}$ corresponds to the special case which is of interest to physics, where $T_{\mu\nu} = 0$, i.e. where no matter is present. By formulating the theory with regard to the metric $\eta_{\mu\nu}$, one takes into account in a simple way the circumstances that, in the absence of matter, the metric field is $g_{\mu\nu} = \eta_{\mu\nu}$. This result is therefore not a proper inference of the theory, but a consequence of taking as point of departure the metric which is used for the field equations. This approach is based on the idea that the gravitational field is responsible only for the deviation of the metric field from flat space, but that the pseudo-Euclidean metric field itself represents the metric of space free of matter, independent of the theory of gravitation.

3.3. *Mach's Principle*

Both formulations of the theory of gravitation — the general theory of relativity and the field theory of gravitation in flat space — produce, for the case of diminishing distribution of matter under the assumptions mentioned, the space-time continuum of the special theory of relativity. Since in the absence of matter the pseudo-Euclidean metric is itself measurable, the difference between a theory of observables and a theory that uses hidden parameters disappears in this case. The pseudo-Euclidean metric field is not defined in these two formulations of the theory, but rather is introduced into the theory in the form of boundary conditions for the gravitational potentials, or as starting metric.

The question of the origin of the metric field $\eta_{\mu\nu}$ is therefore still unresolved. Newton avoided this problem in his mechanics by interpreting motions in which inertial forces participated — for example the rotation of a bucket — as motions relative to absolute space, i.e. by identifying the Galilei-invariant metric field with 'absolute space', which was not further specified. Mach later pointed out that rotation is also only a relative motion between bodies — between the bucket and the fixed stars — and that the assertion that rotation is a motion relative to absolute space, is only correct under the additional assumption that the stars are at rest in space. Since no information is available on this matter, the concept of absolute space is not necessary for the description of possible experiments, and is therefore superfluous for physics. It suffices to observe relative motions between bodies. In practice, accelerated relative motions relative to cosmic masses evoke inertial forces; those relative to terrestrial masses do not.

The special theory of relativity led Einstein to introduce into physics as a new principle the requirement that all events be explainable by observable causes in the sense of field-theoretic interactions. The inertial forces that occur in a rotating bucket must be caused by the rotation relative to cosmic masses, otherwise nothing is observable. The sky of fixed stars must therefore causally induce inertial forces in the bucket when it is in relative motion to it. Einstein set the additional hypothesis that the field which induces these forces, must be the gravitational field of cosmic masses. From this follows the assertion known as Mach's principle: The inertial forces which occur in the accelerated motion of a body with respect to the cosmic mass distribution are caused by the retarding gravitational actions of the cosmic masses. In the more confined language of the general theory of relativity one could say, instead: The metric field of the locally empty space is determined by the gravitational action of the global mass distribution.

Mach's principle should therefore follow from the theory of the gravitational field as a provable proposition. Newton's theory of gravitation does not provide Mach's principle. Einstein, however, presumed that his improved theory of gravitation would give Mach's principle. There are, indeed, some indications for this. For a body in a gravitational field the inertial mass increases proportionately with the gravitational potential. Furthermore, a rotating hollow sphere increases the centrifugal forces of a body rotating in the opposite direction at the center of the sphere. But by far the quantitatively more important contribution to the inertial forces is caused in both these examples by the pseudo-Euclidean metric $\eta_{\mu\nu}$, which is present as a boundary condition for the metric tensor $g_{\mu\nu}$ at spatial infinity.

In the sense of Mach's principle, boundary values for $g_{\mu\nu}$ cannot be regarded as given but must follow from the theory or the distribution of cosmic masses. Einstein therefore supposed that Mach's principle could only be realized in spatially closed, cosmological solutions of the field equations.[55] The question whether Einstein's theory of gravitation can actually provide this result cannot be conclusively answered, even though some interesting positive results in this direction have been given.[56] It is possible that for a complete realization of Mach's principle, a modification of Einstein's gravitation equations or even the introduction of an additional cosmic field is necessary.[57]

We do not wish to enter upon these questions at this point, however, since only a particular aspect of Mach's principle interests us which is of importance in connection with the problem of geometry. For the problem of the foundation of geometry, the essential and new idea derived from

Mach's principle is that the pseudo-Euclidean metric field of the locally empty space can be related to contingent physical sources. So far we defined the metric field by the equations of motion of masses and light rays. Mach's principle declares that the pseudo-Euclidean metric field originates dynamically and does not follow immediately from the method of measurement of lengths and times, i.e. does not follow *a priori*. The metric field of the locally empty space must therefore be strictly distinguished conceptually, from a mathematically defined empirical Euclidean space. The latter, in contrast to the metric field, is determined *a priori* by the geometrical concepts and by the definition of geometrically-rigid bodies used as measuring standards.

4. THE FOUNDATIONS OF A PHYSICAL GEOMETRY

4.1. *Formulation of the Problem*

The question of how the foundations of Euclidean geometry in philosoph-ical apriorism can be reconciled with the results of the general theory of relativity can now be considered. From the *a priori* basis of Euclidean geometry presented in the first section, we can conclude that empirical geometry must also be Euclidean; from the general theory of relativity presented in the second section, on the other hand, it appears that the empirical geometry must be a Riemannian geometry.

On the basis of our considerations of the foundations of Euclidean geo-metry the construction of a geometry in empirical space is fundamentally possible in two ways.

(1) One can start with the fact that geometrical propositions deal with the relations of ideal figures that are defined by rules of construction in pure intuition space. The empirical realization of this pure geometry is then only possible with the aid of measuring standards, whose property of geo-metrical rigidity can only be established with the aid of pure geometry. The validity of the propositions of Euclidean geometry in experience is then established *a priori* under these conditions and no longer depends on contin-gent empirical facts.

(2) As point of departure one can take a certain empirical object, which one defines to be rigid, or a certain empirical line, which one defines to be straight. The properties of the space which one measures with these stan-dards, depend on the specific (in general unknown), properties of the arbitrarily chosen standard.

In the physical theory of gravitation, on the other hand, one starts with the fact that the non-inertial paths of mass points and light rays describe geodesics of a four-dimensional Riemannian manifold. The dynamically defined metric field $g_{\mu\nu}$ is, in the presence of matter, the metric tensor of a Riemannian geometry. In a locally empty region $g_{\mu\nu} = \eta_{\mu\nu}$ is a pseudo-Euclidean metric field. This metric field, however, also has a purely dynamic definition and is therefore conceptually to be strictly distinguished from Euclidean geometry in a mathematical sense. The pseudo-Euclidean metric field as well is of physical origin. It can be assumed that it is created in the sense of Mach's principle by the global distribution of masses in the cosmos. The precise mechanism by which the metric field is produced by the cosmic mass distribution is immaterial for what follows. It is decisive, however, that also the metric field of the locally empty space is caused by empirical facts and is not determined only by the method of experimentation, i.e. *a priori*.

Empirical geometry and the theory of the metric field, therefore, have as yet nothing in common. The geometry is based on well defined constructions of figures and standards in space; the theory of the metric field, on the other hand, is based on the investigation of inertial paths of mass points. The connection between the two theories will become apparent when we consider the empirical geometry actually applied in modern physics, and when we question its foundations (Section 4.2.). Subsequently we shall investigate more precisely the relation of this physical geometry to apriorism. (Section 4.3.).

4.2. *The Foundation of a Physical Geometry within the General Theory of Relativity*

The empirical geometry used in physics is based on the results of the general theory of relativity. The procedure by which this empirical geometry is constructed, however, varies from the two methods of engaging in geometry just mentioned. One commences not with general rules for the construction of measuring instruments and also not with individual empirical bodies, but rather with classes of objects or processes, which are used for the measurement of space. Examples for this are mass points or light rays, or their force-free paths.

The paths of mass points are regarded as geometrically homogeneous. By this identification one is, in principle, able to decide empirically for a sufficiently large number of paths whether, for example, they are geodesics of a Riemannian geometry.[58] The essential advantage of the procedure of using classes of objects as measuring standards, rather than individual objects, is that one does not depend on the particular physical properties of the arbitrarily chosen measuring standard; for the properties of force-free paths or of light rays are physically known. In the use of these processes for the measurement of space any influences exerted on the geometry of space can, therefore, be surveyed, at least in part. From the exact physical properties of the class of objects used as measuring standards one can in general deduce, and state prior to the measurement, certain structures of the geometry, which one empirically measures with these standards.

The physical laws that govern the measuring instruments thus belong to the 'properties' of the measuring standards that give rise to certain propositions on the empirical geometry measured with these standards. In geometrical apriorism, the property of geometrical rigidity of the measuring standard has the consequence that the propositions of Euclidean geometry are necessarily valid in experience. By analogy to this situation one could also designate the empirical-geometrical propositions which derive from the physical properties of the measuring apparatus, as 'synthetic *a priori* judgments'. The essential difference, however, is that in apriorism the judgments follow from the rules of construction of geometrical concepts, whereas here the basis is in physical laws. As an example of such an *a priori* valid proposition, we shall consider three-dimensional space and the possibility of measuring it with a physically rigid measuring standard. The following proposition then applies: "Space measured with sufficiently small, physically rigid measuring rods is Riemannian". Because of the point isotropy of the metric field a sufficiently small, physically rigid body is mechanically freely movable. But since, on the other hand, the Helmholtz proposition brings the measurement of space with sufficiently small, freely movable measuring rods necessarily to a Riemannian geometry, then the empirical space measured with sufficiently small, physically rigid bodies is a Riemannian space.

This example exposes an important condition that must be imposed on the establishing of measuring standards in physics. So far we have tacitly assumed that the establishing of standards by classes of objects is, in the final analysis, arbitrary and that it only has to take place with regard to expediency. That is not the case, however. The geometry measured with real measuring rods is in part determined by the physical laws which govern the measuring instruments. If in addition one requires that these physical laws have an empirically verifiable form, then they can only be formulated if an empirically measurable geometry already exists, to which the propositions of these laws can be referred.[59] The geometrical structure of empirical space, which can be derived from the physical laws that govern the measuring instruments, must therefore agree with the geometrical presuppositions which were already provided as a basis for the formulation of the laws of nature. Thus we must demand that the empirical geometry with its assumptions, i.e. the geometry on which the formulation of the laws of the measuring instruments is based, be consistent.

In the case here discussed, it is the Riemannian character of the metric that on the one hand emerges empirically, but which on the other hand must serve as a basis for the formulation of the laws of nature on their covariant form.

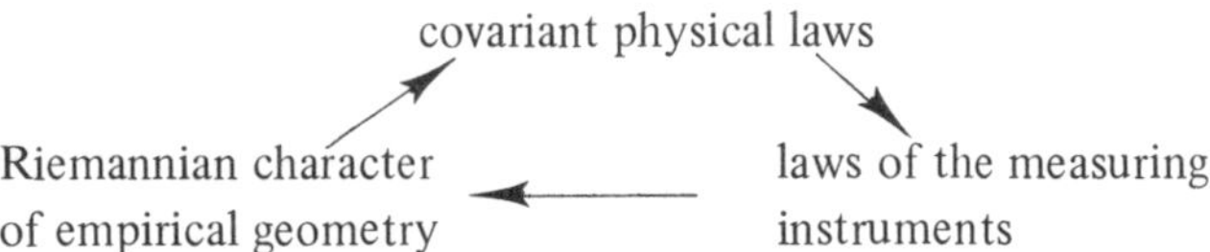

By the requirement of self-consistency, physics acquires a cyclical structure. It is not sufficient to demonstrate the empirical validity of a physical theory, since it is at the same time the theory of the measuring instruments with the aid of which it becomes empirically verifiable. In addition to empirical verification, the proof of the self-consistency of the theory must be provided.

4.3. *The Relation of Physical Geometry to Apriorism*

The procedure just presented in Section 4.2 corresponds exactly to the method by which modern physics deals with empirical geometry. The theory of the metric field in the general theory of relativity forms the basis on which various procedures for the measurement of space were developed and which satisfied the stated condition of self-consistency.[60] It is still an open question as to which relation this procedure enjoys to the concept of empirical geometry which was coined in philosophical apriorism, i.e. whether it is at all possible to designate as geometry in the sense of apriorism a space structure measured with certain classes of objects.

In order to deal with this problem we shall first ask a contrary question, whether it is not possible to set forth an empirical geometry in the sense of apriorism, independent of the described procedure of a physical geometry; an empirical geometry that starts with the construction of its concepts and, using these concepts, defines what constitutes a measuring standard. Thereby a connection between the geometry constructed in this way and the metric field will become apparent, which is characteristic for the problem of the empirical realization of an *a priori* geometry: The physical metric field states to what extent the degrees of freedom of motion of the measuring standards, which are required for the realization of geometry, are available. We shall clarify these relations for Euclidean geometry and the pseudo-Euclidean metric field.

(α) *The empirical realization of Euclidean geometry*

Pure Euclidean geometry takes as a starting point the rules of construction which the definitions relevant for this geometry are defined. With the aid of these concepts, one obtains simple geometrical propositions, which serve as axioms, for the entire geometry. Starting with the geometry introduced in this way, rigid measuring rods are then defined with the aid of which objects can be measured in experience. It is apparent that properties of objects of experience measured in this manner must necessarily obey the propositions of Euclidean geometry.

The measuring instruments, however, must not only be defined but must also be realizable. They are not only the methodological prerequisite for every empirical geometry, but as bodies they are also physical objects. This implies that the conditions which are imposed on the measuring instruments by the definition, must be possible on the basis of physical laws. For the empirical realization of Euclidean geometry unrestricted mobility of geometrically-rigid bodies is necessary. This free mobility is secured in a pseudo-Euclidean metric field, from Mach's principle, by the gravitational action of a sufficiently homogeneous and isotropic distribution of cosmic masses. By the interaction of these two components, the measuring standards defined with the aid of geometry and the maximally homogeneous metric field that makes the real existence of these measuring standards possible, Euclidean geometry is empirically realizable. It is then strictly valid for the arbitrary objects of experience.

Thus we see that the existence of a pseudo-Euclidean metric field is not identical with the validity of Euclidean geometry in empirical space, but that the metric field represents the physical condition for the possibility of an empirical realization of Euclidean geometry. The interaction of the individual factors for an empirical realization of Euclidean geometry may be schematically presented:

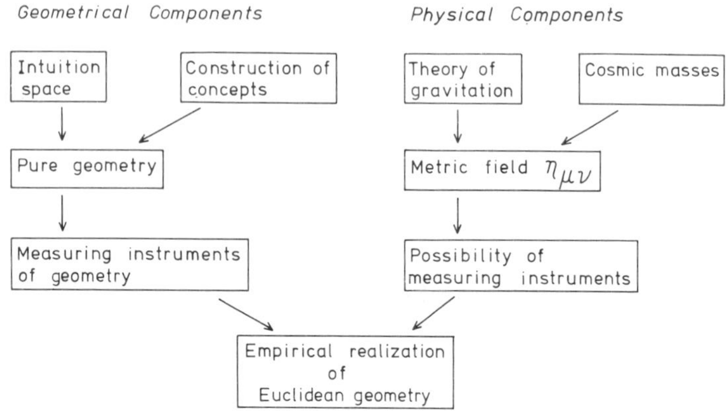

(β) *The realizability of a geometry*

So far we have started with the requirement that a geometry defined by rules of construction must be empirically realizable, i.e. that the measuring standards, which are defined by this geometry and which are to be used for the measuring of reality, should be materially possible. A requirement such as this entirely corresponds to the idea of an *a priori* geometry. Only those geometric rules of construction are established for which one can assume, on the basis of a qualitative prescientific knowledge of experience, that they can be carried out at least in principle.

On the other hand, it is not at all necessary that the measuring standards, with which we are here concerned, can always be produced, especially under adverse circumstances. It is enough that the existence of measuring standards is secured under certain ideal conditions, which in practice can never be realized. So, for example, the isometric transport of a measuring rod is possible only under the condition, never realized in practice, that no local temperature deviations occur. In this instance one will proceed from the assumption that constant temperatures of any desired precision can be produced, at least in principle, and that the existence of measuring standards is therefore possible.

It is then suggested that the gravitational forces, which are the source of the metric field, can be counted among the adverse circumstances which impede practical construction of measuring rods. The connection between geometry and the metric field previously given would then be completely irrelevant for the question of the foundation of geometry. In fact, the foundation of Euclidean geometry, as conceived in philosophical apriorism, appears to proceed from the assumption that one has to imagine the material realization of the measuring instruments in an ideal force-free space, and not in a force field that restricts the mobility of the measuring instruments and subjects them to deformations that depend on position.

According to contemporary physical knowledge there are no objects in nature that are not affected by a gravitational field. Because gravitational forces are long-range, field-free regions do not exist in space. Even in the absence of local matter, the field of the cosmic mass distribution is still present and, provided Mach's principle proves true, this is sufficiently strong to produce inertial forces. The case assumed in philosophical apriorism, that no external forces of any kind act on the measuring standards used, is therefore to be equated with a situation in which no metric field exists that offers resistance to the motion of geometrically-rigid bodies, i.e. in which $g_{\mu\nu} = 0$. If there were no gravitation in the world, i.e. if the gravitational constant $\kappa = 0$, then, with the assumption of Mach's principle, this situation would be realized.

The situation described is, however, purely hypothetical and does not correspond to the given facts of the real world. In the real world, the metric field $g_{\mu\nu}$ is not equal to zero; in the absence of local matter $g_{\mu\nu} = \eta_{\mu\nu}$ to a good approximation. In practice this is of no consequence for the foundation of Euclidean geometry insofar as the field is homogeneous and therefore does not disturb the processor of the geometrical measuring instructions. As for the case of vanishingly small metric, it does not affect the constructions of a measuring geometry.

Presumably here is the reason why most of the philosophical attempts to establish geometry as a science of space failed to see the distinction between an absolutely empty space without a metric field and a space with a pseudo-Euclidean metric field, and simply regarded the empirically given homogeneous space as metric-free space. It seemed reasonable, therefore, to assume an ideal, empty, metric-free space and to specify, in this space, rules of measurement for an empirical geometry. We have seen, however, that the metric field is not to be counted among the arbitrary adverse circumstances that impede measurements, and which can therefore be left out of the considerations; instead it is an ever-present, fundamental property of reality. As long as the metric field is homogeneous, it does not disturb geometry adversely. But since the structure of the metric field depends on the contingent distribution of mass in space, one has to be prepared for situations in which the metric field impairs the basic geometrical constructions or, in fact, makes them impossible.

(γ) *The realization of a Riemannian geometry*

The metric field actually present in the world is not pseudo-Euclidean. In the presence of matter it has, in general, a structure that corresponds to a strictly Riemannian space with a non-vanishing curvature tensor. A metric field of this kind does not allow for unrestricted mobility of bodies necessary for the realization of Euclidean geometry. The empirical realization of Euclidean geometry is therefore no longer possible.

Nevertheless in order to make the measurement of empirical space possible, it would be necessary to find constructions that can be realized in a Riemannian metric field. This is possible in the following manner: In a Riemannian metric field, the 'unrestricted mobility' used so far is no longer available, but rather only a 'restricted mobility'. Within the precision of physical measurements, objects of finite extension are freely movable in a bounded region, that is, sufficiently small bodies are movable without restriction in the entire space. For the construction of a geometry, it is not necessary to know the precise details of the metric field, since the qualitative property of restricted mobility suffices.

Because of the free mobility of finite bodies in a restricted region, it is possible to realize a Euclidean geometry in the vicinity of every point, the precision of which depends inversely on the size of this neighbourhood. In the limit, for a sufficiently small neighbourhood, a Euclidean space belongs to every point of space. If one regards these Euclidean spaces as tangential spaces of a Riemannian manifold, then a Riemannian geometry has been empirically realized by the totality of Euclidean geometries realized at every point. One can then use the Riemannian geometry defined in this way to define sufficiently small, geometrically-rigid measuring standards that have unrestricted mobility, as required, and thus to measure actual space. The empirical space measured with these standards is then necessarily a Riemannian space.

The relation between the metric field and empirical geometry can be defined as shown in Table II.1. Euclidean geometry is therefore in general not realizable; an empirical Riemannian geometry, on the other hand, is always possible.

TABLE II.1

Metric field	Property	Euclidean geometry realizable	Geometry
Pseudo-Euclidean	unrestricted mobility	unrestricted	Euclidean
Riemannian	restricted mobility	locally	Riemannian

(δ) *Objections raised by apriorism*

If one starts from a geometry whose concepts are defined by rules of construction, then the measuring standards of an empirical geometry are defined by geometrical properties. The question whether the ideal requirements on which the construction of measuring standards are based are meterially realizable has so far been answered in the following sense: The physically defined metric field $g_{\mu\nu}$ decides whether a certain geometry is empirically realizable or not.[61] The metric field is the physical condition for the possibility of an empirical geometry.

The program for the construction of the exact sciences[62], conceived in apriorism, looks quite different methodologically. Starting from prescientific empirical knowledge, certain qualitative features of the natural environment are brought to prominence. These assure that some primitive procedures are possible. To these belongs the fact that distinguishable, discrete things are found in the world, whereby the possibility of an empirical arithmetic is given. Dingler's grinding procedure for producing plane surfaces is also based on the prescientific knowledge that bodies, e.g. stones, exist which can be mutually ground. On the basis of this qualitative cognition of nature, one can therefore construct arithmetic and geometry and thereby provide the mathematical foundations for an exact natural science. Exact science is established if, in addition to mathematics, some empirical facts are taken into account, as for example the empirical properties of gravitational forces. But only on the basis of such a mathematical theory of gravitation is it possible to construct an exact theory of the dynamic metric field.

If one takes note of this linear methodological order, then it is impossible to base the geometrical constructions on the theory of the metric field, as we have done so far. For the theory of the metric field is only possible within the framework of mathematical physics, and this presupposes arithmetic and geometry in its construction. The question whether the fundamental geometrical constructions can be realized empirically must take its direction from prescientific empirical knowledge. This orientation, however, will be executed under the view that the ideal rules of construction for geometrical concepts can at least approximately be carried out. Thus, within the framework of apriorism the question of empirical realizability is also drawn into consideration; the possibilities of realization, however, are decided on the basis of a qualitative knowledge of nature, and not on the basis of the theory of the metric field.

The methodological justification of the idea, outlined above, that the metric field is to be regarded as the physical conditions for the possibility of an empirical geometry, obviously depends on what is designated as prescientific experience. It is undoubtedly significant that it is a matter of qualitative knowledge, not yet formulated in mathematical concepts and relations. Apart from this formal property, the material content of knowledge is not decisive for its prescientific character.

The prescientific knowledge, on which philosophical expositions are commonly based, derives from the day-to-day experience with objects and processes of the natural human environment. It is not, however, representative of reality in general. The basis for this is that even the qualitative properties of nature that surrounds us cannot be determined independently of the dimension of the objects under consideration. The objects of the human environment are remarkable because of their particularly simple qualitative properties: they are so large that quantum phenomena are practically not observable, i.e. the concept of thing cannot be meaningfully used for the ordering of experience.[64] On the other hand, they are so small that for all practical purposes the restriction of free mobility of bodies, important for cosmic dimensions, is not yet noticeable.

It is therefore not very meaningful to base the forming of concepts on the prescientific experience, which is familiar to us from the natural environment of mankind. For the proper domain of application of these concepts is not to be the scientific formulation of this prescientific experience, but the physical formulation of phenomena of cosmic extent. Only for phenomena of this kind is the difference between Euclidean and Riemannian geometry relevant. It would therefore be reasonable to derive geometrical concepts, which are to be applicable not only to the human environment but also to cosmic phenomena, from a prescientific experience that embraces this cosmic environment.

There still remains the question of how one comes to possess such a cosmic, prescientific experience. The following possibility presents itself: Starting with Euclidean geometry, which can be *a priori* justified, one investigates the field theory of gravitation in flat space, as we have done in Section 4.2 of the 'theory of the metric field'. From this theory one determines the experimentally observable metric field $g_{\mu\nu}$. From the complete theory of the metric field, thus obtained, the desired, purely qualitative properties of the metric field can be determined. One of these is the property that finite, geometrically-rigid bodies are freely movable only in a bounded region; only sufficiently small bodies therefore have complete free mobility. The property of restricted mobility suffices to serve as a basis for the formation of geometrical concepts. The qualitative property of restricted mobility is a constraint on the unrestricted mobility familiar from man's experience in his environment. As we have seen previously, it leads, not to a Euclidean, but to a Riemannian geometry of empirical space.

In this manner the previously stated procedure of giving direction to the construction of the fundamental geometrical concepts by recourse to the circumstances of the metric field can be comprehended as a geometry in the *a priori* sense. Not a mathematically formulated theory of the metric field is offered as a basis for the fundamental geometrical constructions, but rather only the qualitative properties of the restricted or unrestricted mobility of bodies. Knowledge of these properties corresponds to prescientific experience. On the basis of unrestricted mobility, it is possible to construct Euclidean geometry as an *a priori* theory, which is necessarily valid for all objects of the prescientific domain of experience. Similarly, Riemannian geometry can be established on the basis of restricted mobility. It too is an *a priori* theory in the sense explained, and is necessarily valid for all objects of experience.

5. SUMMARY

The investigations of this chapter have shown that Euclidean geometry can be comprehended as a theory that has its point of departure in the construction of its fundamental concepts, and whose axioms necessarily follow from the procedures for construction of the concepts. This theory, which deals with ideal figures in intuition space, is applicable under certain conditions to empirical space. One of these conditions is that empirical things have the property of unrestricted mobility. The propositions of Euclidean geometry are then strictly valid for all objects of experience.

In this procedure, methodological order in the construction of a scientific geometry is assured. Since experience is already possible prior to having a scientific geometry, one can infer from this prescientific experience that empirical objects have the pregeometric property of unrestricted mobility. Pure geometry constructed in intuition space can then, with the aid of well-defined measuring standards, be transposed to the empirical space.

In the general theory of relativity, the mobility of objects is investigated in the framework of a physical theory. It becomes apparent that this mobility is a special property of the metric field. Apart from seldom realized special cases, the empirically given metric field does not, however, assure the unrestricted mobility so far required, but only the less stringent property of restricted mobility. The qualitative prescientific property of the metric field cannot be perceived directly, but only by way of a physical theory. This circumstance is engendered by the fact that this property only acquires importance in dimensions that significantly exceed those of the immediate world of human experience.

Since Euclidean geometry can thus not be realized empirically, two possibilities present themselves for a methodologically correct construction of the general theory of relativity.

(1) One relinquishes the idea of realizing an *a priori* geometry empirically, and instead one starts with light rays and inertial paths of mass points, basing the mathematical formulation of the theory on the geometry realized by these processes. This physically defined geometry is shown to be a Riemannian geometry. By thus avoiding the use of the empirically unrealizable Euclidean geometry in the formal construction of the theory, the general theory of relativity becomes a theory which contains only observables, and which describes nature as it is revealed when measured with

certain real measuring standards (e.g. light rays). Since light rays and paths of mass points too are objects of the theory, it must have the property of self-consistency. There exists no connection between the Riemannian geometry realized in this theory and the notion of a scientific geometry conceived in philosophical apriorism.

(2) One starts with the *a priori* Euclidean geometry and makes use of it as a mathematical framework for the theory of gravitation, without demanding its empirical realization. The metric field calculated in this theory then has the same structure as in the theory of observable magnitudes. For this reason equivalent propositions on the mobility of objects are obtained. The Euclidean metric, on which the theory is based, represents a system of hidden parameters that cannot be observed in any possible experiment.

The metric field determined in this manner has the simple qualitative property of restricted mobility. If one interprets this property as a pregeometric property, which originates from prescientific empirical knowledge, then one can ask what possibilities for the construction of an *a priori* geometry still exist in this case? As was shown, a Euclidean geometry is realizable only locally under this prescientific condition, whereas a Riemannian geometry obtains in the entire space. One may therefore regard the Riemannian geometry constructed in this way as the realization of an *a priori* geometry, in the same sense that would be considered appropriate for the realization of Euclidean theory in a homogeneous metric field.

These two possibilities of interpretation of the Riemannian geometry used in the general theory of relativity are once more schematically summarized in the Table II.2. On the left-hand side the two possible formulations of the theory of gravitation are stated, on the right-hand side the consequent geometrical theories.

TABLE II.2

Physics	Geometry
1. Theory of observables. Self-consistency of the theory with the measuring instruments	Geometry of light rays and mass-point paths ⇒ Riemannian geometry
2. Euclidean geometry as mathematical framework; theory of hidden parameters	Restricted mobility as pregeometric knowledge ⇒ construction of an *a priori* Riemannian geometry

NOTES

[1] Plato, *Republic VI*, 510c–511b.

[2] *Ibid.*, 526a–527a.

[3] Plato, *Meno*, 82b–85c.

[4] Plato, *Timaeus*, 50B.

[5] Immanuel Kant, *Kritik der reinen Vernunft*, 1781(A), 1787(B); *Critique of Pure Reason* (transl. by N. K. Smith), Macmillan, London, 1929, p. (B)4, p. 44.

[6] Immanuel Kant, *Prolegomena*; Lucas *Prolegomena* (transl. by P. G. Lucas), Manchester Univ. Press, Manchester, 1953, p. (A)39/40, p. 29, also Kant, *Critique*, p. (B)128.

[7] Kant, *Critique*, p. (B)40, p. 70.

[8] *Ibid.*, p. (B)40, p. 70.

[9] *Ibid.*, p. (B)40, p. 70.

[10] *Ibid.*, p. (B)199, p. 195.

[11] *Ibid.*, p. (B)761, p. 589.

[12] In Kant's usage, 'principle of contradiction' refers to the system of propositions of logic.

[13] Kant, *Prolegomena*, p. (A)28, p. 18.

[14] Kant, *Critique*, p. (B)741, p. 577.

[15] *Ibid.*, p. (B)741, (B)742, p. 577.

[16] *Ibid.*, p. (B)40, p. 69.

[17] *Ibid.*, p. (B)204, p. 198.

[18] *Ibid.*, p. (B)204, p. 199.

[19] *Ibid.*, p. (B)743, p. 578.

[20] *Ibid.*, p. (B)744, p. 578.

[21] *Ibid.*, p. (B)761, p. 589.

[22] Because of the quantitative character of all figures that can be constructed in space, the intuition space can presumably also be rendered metrical. The uncertainty of this comment is caused by the difficulty of developing the mathematically precise property of metricizability from the rather qualitative arguments of Kant.

[23] *Ibid.*, p. (B)198, p. 194.

[24] *Ibid.*, p. (A)162, p. 196.

[25] *Ibid.*, p. (B)202, p. 196.

[26] E. Richter has pointed out ('Kant und die Frage nach den Einheitsfunktionen der Erkenntnis', Manuskript 1965) that these considerations could receive further support from the transcendental deduction of the categories.

[27] *Ibid.*, p. (B)206, p. 200.

[28] Kant, *Prolegomena*, p. (A)62, p. 44.

[29] The thesis that the Euclidicity of space is constitutive for experience, was advanced by Heinrich Scholz, since in that case the assertion, repeatedly made by Kant, that Euclidean geometry is strictly valid for experience can be easily understood. (H. Scholz, *Mathesis Universalis*, Basel 1961, p. 204). I believe, however, that one must, at this point, bear in mind the difference, expressly emphasized by Kant, between mathematical and transcendental-philosophical propositions. Whereas the principles of pure reason follow from the conditions of the possibility of experience, the truth of mathematical propositions is based on the construction of its concepts and on immediate intuition.

[30] Kant, *Prolegomena*, p. (A)29, p. 20.

[31] Kant, *Critique*, p. (B)65, p. 86.

[32] *Ibid.*, p. (B)745, p. 579.

[33] Ernst Cassirer has advanced the thesis that Kant did not at all have Euclidean geometry in mind when speaking of the empirical validity of geometry. As justification, Cassirer correctly notes that Euclidicity is not essential for experience, and that empirical space is necessarily only a topological, metricizable space. Cassirer is mistaken, however, in asserting that empirical space then already is a Riemannian space, and in believing that he has eliminated by these considerations the differences between the empirical Riemannian geometry of the general theory of relativity and the Kantian apriorism of Euclidean geometry. (E. Cassirer, *Zur Einsteinschen Relativitätstheorie*, Berlin 1921, p. 85; 101 [see English transl. by M. C. and W. Swaley, *Substance and Function, and Einstein's Theory of Relativity*, Open Court, Chicago, n.d.]). Apart from the erroneous assumption that space as form of phenomena is already a Riemannian space, it has been repeatedly emphasized by Kant that geometry comes about in space only by the construction of its concepts, but that this geometry is applicable to experience. This supports the idea that Kant clearly meant Euclidean geometry.

[34] The triangle was formed by the Inselberg, the Brocken and the Hohen Hagen. The sides of this triangle are 69, 85 and 107 km.

[35] H. v. Helmholtz, 'Über die Tatsachen, die der Geometrie zum Grunde liegen', in *Reden und Vorträge II*, Braunschweig 1896, p. 1 and 'Über den Ursprung und die Bedeutung der geometrischen Axiome', in *Wissenschaftliche Abhandlungen II*, Leipzig 1883, p. 168. [See English transl. of these essays by M. Lowe in *Epistemological Writings of Hermann von Helmholtz* (ed. by R. S. Cohen, Y. Elkana and M. Lowe), *Boston Studies in the Philosophy of Science*, volume 37, Dordrecht and Boston 1975.]

[36] For a proof see, for example, D. Laugwitz, *Differentialgeometrie*, Stuttgart 1960, p. 145f.

[37] For a coordinate-free formulation of these relations see H. Busemann, *The Geometry of Geodesics*, New York 1955.

[38] Instead of defining a rigid measuring standard, with the aid of which the straightness of a line can be determined, one can use the opposite procedure of defining a particular empirical surface, e.g. the earth's surface, as flat. Measuring standards and the geometry can then be derived from this definition. It is obvious that this method yields no experimental propositions about space; measurements depend on the behaviour of the surface just defined. A detailed discussion of this procedure is found in R. Carnap, 'Über den Raum', *Kantstudien* **56** (1922), p. 47.

[39] The idea that physics could be changed and Euclidean geometry retained, is already found in Helmholtz, 'Über den Ursprung . . .', *ibid.*, p. 29, and is extended by H. Poincaré, *Wissenschaft und Hypothese*, Leipzig 1914. [English transl. *Science and Hypothesis*.] R. Carnap, *ibid.*, also emphasizes this possibility and concludes that the establishing of standards is completely arbitrary.

[40] H. Poincaré, *Wissenschaft und Hypothese*, German edition, Leipzig 1914, p. 51, 52.

[41] D. Hilbert, *Über die Grundlagen der Geometrie*, Leipzig 1899.

[42] It is to be noted, however, that the straightness of a line, for example, cannot be determined for an individual line by the use of Euclidean axioms; only for a sufficiently large number of lines, recognized as similar by criteria other than geometrical, is this possible. For expediency such test is carried out not with the aid of synthetic

geometry, but with an analytical formulation. G. Holland, Dissertation, Hamburg 1964, examines more closely how it is possible in an analytical formulation of geometry to determine whether, for example, experimentally given orbits of mass points are straight lines.

[43] H. v. Helmholtz, 'Über die Tatsachen, die der Geometrie zum Grunde liegen', *ibid.*, p. 618.

[44] H. v. Helmholtz, 'Über den Ursprung und die Bedeutung der geometrischen Axiome', *ibid.*, p. 30.

[45] P. Lorenzen, 'Das Begründungsproblem der Geometrie', *Phil. Natural*, VI, 415 (1961).

[46] H. v. Helmholtz, 'Über den Ursprung und die Bedeutung der geometrischen Axiome', *ibid.*, p. 30.

[47] H. Dingler, *Über Geschichte und Wesen des Experiments*, Munich, 1952.

[48] If one requires that operative definitions be carried out with homogeneity relations, then only the possibility of geometries of Riemannian spherical spaces remains. The concept of length only becomes definable with the aid of congruence. If one therefore adds the further requirement that no specific length appear in the definition, then it can be shown that only Euclidean geometry can have an operative basis.

[49] Details of the procedures of measurement can, for example, be found in: C. Møller, *The Theory of Relativity*, Oxford 1952; W. Kundt and B. Hoffmann in *Recent Developments in General Relativity*, p. 303. PWN, Warsaw, 1962; R. F. Marzke and J. A. Wheeler, in *Gravitation and Relativity,* (ed. by H. Y. Chiu and W. F. Hoffman) W. A. Benjamin, New York, 1964, p. 40.

[50] In accordance with the literature written in English, we use here the term 'metric field', although it anticipates the geometrical interpretation of this field due to the prefix 'metric'. In the German edition of this book we have used the more neutral form 'Führungsfeld' (guiding field) which was introduced by Herman Weyl (1922) in order to express the primarily dynamical character of this field. Since we do not use this term here, the 'metric field' must be clearly distinguished from its geometrical interpretation as a 'metric tensor'.

[51] P. Mittelstaedt and J. B. Barbour, *Z. Physik* **203** (1967), 82.

[52] On this question see F. Rohrlich, *Annals of Physics* **22** (1963), 165.

[53] See especially S. Gupta, *Proc. Roy. Soc.* **A65** (1952), 608 and *Rev. Mod. Phys.* **29** (1957), 334; W. Thirring, *Ann. Phys.* **16** (1961), 96. W. Wyss, *Helv. Phys. Acta* **38** (1965), 469; J. B. Barbour, *Dissertation*, Cologne (1967).

[54] The local equivalence of the Lorentz-invariant gravitation theory and Einstein's theory is apparent from the reasons given. The question of global equivalence, however, requires a separate discussion. As in Newtonian cosmology (see O. Heckmann and E. Schücking, *Handbuch der Physik* LIII, Berlin 1959, p. 489) so also in the Lorentz-invariant theory of gravitation the investigation of the system of equations (consisting of field equations and equations of motion) that describes the motion of infinitely extended cosmic matter under the action of its own gravitational field leads to different solutions for the density and the velocity field of the cosmic matter, which correspond to the possible cosmological models. Together with the special empirical world postulate of homogeneity and isotropy of the cosmos, one thus arrives at solutions that, after transposition to Riemannian space, agree with Friedmann's solutions of Einstein's theory. (P. Mittelstaedt, *Z. Phys.* **211**(1968), 271.

[55] A. Einstein, *The Meaning of Relativity*, Princeton 1955; see also H. Hönl and H.

Dehnen, *Z. Physik* **166** (1962), 544; *Ann. d. Physik* **11** (1963), 201; **14** (1964), 271.

[56] F. Gürsey, *Ann. Phys.* **24** (1963), 211.

[57] See for example: F. Hoyle, *Monthly Notices Roy Astron. Soc.* **108** (1948), 372; F. Hoyle and J. V. Narlikov, *Proc. Roy. Soc.* **A270** (1962) 334; H. Yilmaz, *Phys. Rev.* **111** (1958), 1417; C. Brans and R. H. Dicke, *Phys. Rev.* **124** (1961), 925.

[58] How this is to be done in detail is discussed in G. Holland, Dissertation, Hamburg 1964.

[59] As shown above, the theory of the metric field can in principle be formulated in a pseudo-Euclidean metric, which, however, is not observable in any possible experiment. It is therefore impossible to verify empirically the theory in this formulation.

[60] Refer to the methods of measurement given by C. Møller, *Theory of Relativity* or by J. L. Synge, *Relativity: The General Theory*, Amsterdam 1960.

[61] The fact that even in a flat metric field an ideal plane can never be realized because of the atomic structure of matter is immaterial to our inquiry and therefore is neglected.

[62] See for example P. Lorenzen, 'Wie ist die Objektivität der Physik möglich', in *Argumentationen: Festschrift für Josef König*, Göttingen 1965.

[63] It is not necessary here to express the dimensions of objects by geometrical magnitudes; prior to formulating a geometry this is not possible. Atomic, classical, and cosmic bodies could be distinguished, for example, by the number of atoms of which they consist; the order of magnitude being 10, 10^{25}, and 10^{50}, respectively.

[64] See Chapter IV.

CHAPTER III

THE QUANTUM-MECHANICAL
MEASUREMENT PROCESS

The observation of atomic systems is only possible with measuring devices that actually occur in nature, i.e. that are themselves constructed of elementary particles. The theory that describes the events of nature, as shown by such measuring instruments, is the quantum theory. The inherent influence that measuring devices exert on observational results is determined by the physical laws of the measuring instruments, i.e. by quantum theory itself. The theory of the quantum-mechanical measuring process deals with these relations.

1. THE UNCERTAINTY RELATION

In non-relativistic classical mechanics, the space-time behaviour of a physical system is described by giving the values of all possible observables $A_1, A_2, A_3 \ldots$, i.e. of all measurable properties of the system. We shall denote the state of an object as its characterization by the ensemble $A_1, A_2 \ldots$ of all possible measurement values. Knowledge of the state then permits exact prediction of the changes with time of all properties of the system. Classical mechanics has shown, however, that to specify the state of a system it is by no means necessary to ascertain all and therefore infinitely many properties, but that the state is already uniquely determined by the position x and the momentum p, or by any other pair of canonically conjugate variables. The time-dependent changes of x and p are then described by Hamilton's equations.

Strictly speaking, the state (x, p) of a system would be physically meaningful only if it were established how the state could be experimentally ascertained. In the framework of classical mechanics this question was as little discussed as, for example, the problem of space and time measurement before the theory of relativity was established. A procedure of state assignment compatible with the principles of classical mechanics would use the idealizing assumption of measuring instruments, suitable for the measurement of the quantities x and p, whose precision could be increased indefinitely. For example, one could imagine measurements of position to be carried out with light of sufficiently short wavelength λ. From classical mechanics no objection can be made to such idealization, since no laws of nature appear in this theory which would set an objective upper limit independent of the observer to the precision of measurement.

The situation is changed fundamentally if the results of atomic physics, not part of mechanics, are taken into consideration: matter is made up of elementary particles, and even electromagnetic radiation has this quantized nature. Especially the latter discovery, which goes back to Planck and Einstein, has been of decisive significance for further developments. Light waves, which can be electrodynamically characterized by their frequency ν or their wavelength λ, are correlated with light quanta, photons, to which one ascribes a momentum p and an energy E, as for other elementary particles. The connection between these partial definitions is given by the two relations:

$$E = h\nu \qquad p = h/\lambda$$

where h is Planck's constant. These relations state that light of a certain frequency ν must have at least an energy $h\nu$ and a momentum h/λ. Smaller units of light do not exist. As long as one did not know of the existence of the quantum of action $h > 0$, however, one could assume that arbitrarily small units of a given light do in fact exist. This realization has far-reaching consequences for a discussion of the measuring processes for the quantities p and x. As a simple case we shall discuss the measurement of the position of an elementary particle, an electron for example, by the use of a microscope.

Suppose the light used has frequency ν and wavelength λ. Then for an angle of observation ϵ, the wave theory of light gives an uncertainty of the position coordinate in the x direction of

$$\Delta x \sim \frac{\lambda}{\sin \epsilon} \; .$$

Thus, the shorter the wavelength of the light, the more precise the measurement of position will be.

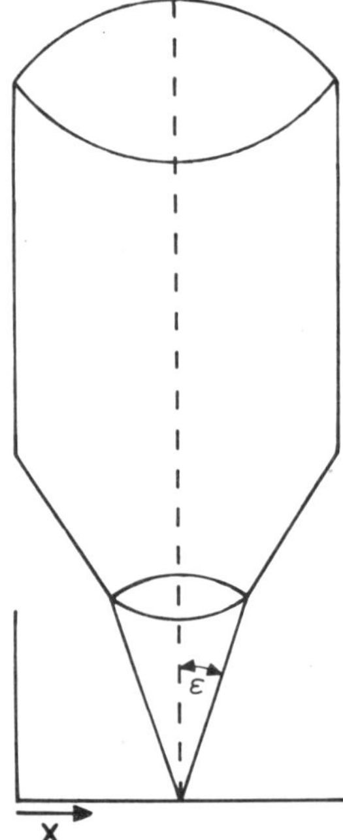

Fig. 7. Uncertainty in the measurement of position.

It is immediately apparent, however, that in such a measurement the momentum information, is lost at least in part even if it were known precisely prior to the measurement. The measurement requires at least one quantum of light. Because of the quantum structure of light, it has an energy $E = h\nu$ and a momentum $p = h/\lambda$. The observed electron therefore experiences a Compton-recoil from the quantum of light. The direction of the photon is not precisely known, however, since it can vary within the angle ϵ. For the momentum transfer in the x-direction the uncertainty

$$\Delta p_x = \frac{h}{\lambda} \sin \epsilon$$

results. Together with the relation for the uncertainty of position, derived above, it follows that:

$$\Delta x \, \Delta p \sim h.$$

This relation, known as the Heisenberg uncertainty relation, states that it is not possible to measure p precisely and x precisely at the same time. If one increases the precision of the x-measurement by, for example, using light of shorter wavelength, one thereby impairs the momentum information. Conversely, the precision of the x-measurement is reduced if one decreases the Compton-recoil by using light of greater wavelength.

We have investigated a particular example for which the uncertainty relation could be derived because of the quantum nature of light. But this relation is also valid when the measurement is carried out, not with light, but with other material particles. One merely has to consider that the equations

$$E = h\nu \qquad p = h/\lambda$$

can, following DeBroglie, be read so that a particle with the values E and p is assigned a matter wave of frequency ν and wavelength λ.

Using the wave nature of matter and the quantum nature of light it has been shown in a large number of mental experiments that it is never possible to measure more precisely than is permitted by the uncertainty relation. Since matter and radiation are the only realities occurring in nature — disregarding gravitational fields — it follows that it is in general impossible to use measurements to determine the state of a system more precisely than is permitted by the limitation of the uncertainty relation.

In particular, it is clear that for the example of the microscope the uncertainty in the momentum cannot be avoided by measuring the recoil of the photon on the microscope. This recoil would have to be measured with a second measuring instrument, which, ultimately, also consists of matter and radiation and thus is subject to the same uncertainty relation for position and momentum. The difficulty therefore is displaced, but not eliminated.

The physical reason for this new situation is that Planck's constant h differs from zero. By comparison to effects occurring in macroscopic physics, $h = 6.623 \times 10^{-27}$ erg sec is of course very small. The fact that $h > 0$ could, therefore, be neglected in classical physics.

Not until one considers atomic processes, in which energies are involved that are comparable to $h\nu$ for small values of ν, does it become necessary to take into account the finite value of h. The circumstances are therefore much as in the theory of relativity, where limitations on the possibility of measurement by the finiteness of the speed of light were also only taken into account when phenomena at very high speeds were investigated.

The highly schematic experimental arrangement for the measurement of position, used so far, is still not quite satisfactory. The perturbation that the measuring instrument exerts on the object is, ultimately, caused by an interaction. The question therefore arises, whether it is possible to calculate the perturbation by allowing for this interaction in the calculation or, if this is not possible, to show why this possibility is excluded.

In order to be able to deal with this problem, we must go beyond the methodology of the semi-classical presentation used so far. The uncertainty relation was derived entirely within the framework of classical physics, only the two relations $E = h\nu$ and $p = h/\lambda$, which relate particles and waves, extended beyond this. To treat the interaction between the measuring device and the object in a physically satisfactory way, it is necessary to use the entire theory of quantum systems, i.e. quantum theory. Only if it is shown in this theory that it is not possible, either experimentally or by calculation, to make statements about the object system beyond the limits of knowledge asserted in the uncertainty relation, may the previous representation of the measuring process be regarded as satisfactory. In the following section we shall therefore briefly consider the essential features of quantum theory, so that a complete description of the measuring process, making use of quantum theory, can be given.

2. QUANTUM THEORY

The considerations of the preceding section are verified within the framework of quantum theory, i.e. it will be shown that it is impossible to determine precisely the values of all observables of a physical system by using real existing measuring instruments. Because of this circumstance it appears reasonable – as in the theory of relativity – to base the theory not on unobserved and, as we know, unobservable objects, but on natural phenomena as revealed by realizable measuring devices. The theory that accomplishes this in the domain of atomic processes is quantum mechanics. Its propositions relate to the results of measurements on quantum-physical systems. Within its range of application this theory has been verified in every way by the results.

In principle there is the alternative possibility of describing events of nature by retaining the framework of the so-called classical physics, in which the influence of the measuring devices on the results of measurement are neglected. In doing so, one would relinquish the possibility of an empirical verification of the theory. A description of experimental results could then only be achieved by supplementing the theory with an exact theory of the measuring process. In the actual development of the physics of atomic processes, however, this possibility played no significant role. The few attempts of this kind – the theory of David Bohm, in particular, should be mentioned – contain, and that may well be their decisive weakness, an enormous arbitrariness in the description of unobservable quantities, the so-called 'hidden parameters', since any possibility of empirical verification is eliminated. We shall therefore consider these attempts only briefly in connection with the problem of causality (Chapter V).

In the framework of quantum theory, a physical system S (an elementary particle, an atom, a molecule, etc.) does not simultaneously possess all properties E_{A_1}, E_{A_2}, E_{A_3} ..., but only some of them. We shall call these the 'objective' properties. The reason for the designation 'objective' will become apparent in the course of this and the following chapters. Briefly, it is because only an objective property can reasonably be associated with an object system; it is not necessarily decided, however, whether it occurs or not. With regard to system S, these objective properties are commensurable, i.e. they can be measured on the object in any sequence, without changing their value. A knowledge of all objective properties is the most complete information that can be obtained on a system. It is identified by the state $|\varphi\rangle$ of the system, or by the property E_φ that corresponds to the occurrence of state $|\varphi\rangle$. A knowledge of the state then makes it possible to predict the measured values of all objective properties.

In quantum theory the states $|\varphi\rangle$, $|\psi\rangle$, $|\chi\rangle$ are represented by vectors in Hilbert space; measurable quantities (observables) by Hermitian operators $A, B, C \ldots$. If an observable A is objective with regard to the state $|\varphi\rangle$, then $|\varphi\rangle$ is an eigenstate of A. We shall restrict our consideration to operators with discrete spectra. The eigenvalue A_i then gives the value of A in the state that we correspondingly designate $|A_i\rangle$. The equation $A|A_i\rangle = A_i|A_i\rangle$ then holds. In the following we shall always assume that the measurement values A_i are mutually exclusive, i.e. the states are orthogonal and satisfy the condition $\langle A_i|A_k\rangle = \delta_{ik}$ for normalized states $|A_i\rangle$. In addition we assume that the states $|A_i\rangle$ span the entire Hilbert space, i.e. that the states $|A_i\rangle$ are complete; this finds expression in the condition $\sum_i |A_i\rangle \langle A_i| = 1$. If two observables are commensurable, then their corresponding operators commute.

The measurement quantity that describes the occurrence of the state $|\varphi\rangle$ is represented by a particular Hermitian operator, the projection operator $P_\varphi = |\varphi\rangle \langle \varphi|$ that projects on the vector $|\varphi\rangle$. It has the eigenvalue 1 whenever $|\varphi\rangle$ occurs, and the eigenvalue 0 when $|\varphi\rangle$ does not occur. Because of the orthogonality of the eigenstates $|\varphi_i\rangle$ of P_φ, $P_\varphi|\varphi_i\rangle = |\varphi_i\rangle$ if $|\varphi_i\rangle = |\varphi\rangle$, and $P_\varphi|\varphi_i\rangle = 0$ if $|\varphi_i\rangle$ is orthogonal to $|\varphi\rangle$. Since orthogonal states are mutually exclusive, the occurrence of a state $|\varphi_i\rangle$ perpendicular to $|\varphi\rangle$ indicates that $|\varphi\rangle$ is not present.

Observables that have only the eigenvalues 0 and 1 and that are described by projection operators P_E we shall call properties E. If P_E has the eigenvalue 1, then the corresponding property E is present; if it has the value 0, it is not. Occasionally we shall say instead that the property not-E, represented symbolically by $\neg E$, is present. Because of the completeness of the eigenstates $|\varphi_i\rangle$ of P_E, the projection operator associated with the property $\neg E$ is given by $P_{\neg E} = 1 - P_E$. It not only projects on a certain state, but on the Hilbert subspace that is spanned by all the states that belong to the eigenvalue 0.

In the following considerations it will frequently be useful to investigate, instead of arbitrary observables, properties for which the measuring process is reduced to a simple yes or no decision. This does not signify a limitation of generality, for the measurement of an observable A with the eigenvectors $|A_i>$ and measurement values A_1, A_2, ... can always be traced to the measurement of properties. For this purpose it is only necessary to measure the operators $P_{A_i} = |A_i><A_i|$ that project on the states $|A_i>$, or rather the corresponding observables E_{A_i}. Because of the assumed orthogonality and completeness of the states $|A_i>$ these measurements are equivalent to the measurement of the observable A. In particularly simple cases the reduction of the measurement to a yes or no decision comes about from the fact that the Hilbert space is two-dimensional. A pertinent example is the measurement of electron spin in the Stern-Gerlach experiment. We shall therefore frequently refer to this and similar experiments for the elucidation of the measuring process. The conclusion drawn from such experiments can, of course, also be derived from more complicated experiments.

It occasionally happens that one does not know exactly in which state a system is to be found; one only knows that it is in state $|\varphi_1>$ with probability w_1, in $|\varphi_2>$ with probability w_2, etc. In this case one speaks of the system as a mixture. A mixture can no longer be characterized by a projection operator. Instead we have the statistical operator

$$W_\varphi = \sum_{i=1}^{\infty} w_i P_{\varphi_i} = \sum_{i=1}^{\infty} w_i |\varphi_i><\varphi_i|,$$

which is a summation of the respective projection operators.

The properties of state $|\varphi>$ that can be measured and calculated simultaneously and which we have called objective represent a special case seldom realized. For a particular property E one cannot, in general, assert that it is attributable to the system in state $|\varphi>$, nor that this is not the case. We designate these properties as *not-objective* with regard to $|\varphi>$.

Yet even these properties are measurable in a certain sense. The state $|\varphi>$ of the system is changed by the measuring process — the measurement of the observable A, for example — in a manner such that with regard to the new state $|A_i>$ the observable A represents an objective quantity; its value A_i can therefore be calculated or measured in the new state. Under certain circumstances an infinite number of eigenstates $|A_i>$ belong to the operator A. But to the question of which of these occurs in the measuring process, quantum theory gives only an incomplete answer. It merely provides probability assertions which relate to the outcome of very many similar experiments. Thus for the probability $w_\varphi(A_i)$ of observing the state $|A_i>$ in the measurement of the observable A of a system in state $|\varphi\rangle$, quantum theory provides the expression:

$$w_\varphi(A_i) = |\langle\varphi|A_i\rangle|^2 = \langle A_i|P_\varphi|A_i\rangle$$

In general, different properties E_A, E_B, E_C will only objectively occur in different states $|A_i\rangle$, $|B_i\rangle$, $|C_i\rangle$. For this reason the measurement of property B will invalidate any preceding measurement of property A, if the B-measurement is associated with a change of state. Such properties can then not be 'simultaneously' related to the system and are therefore called incommensurable.

Two quantities A and B can be incommensurable to different degrees. If, for example, observable A is known, then the probability distribution $w_A(B)$ for the B-values can have a more or less well-defined maximum at a value. Moreover, if the maximum is represented by a δ-function, then B is commensurable with A. On the other hand, if $w_A(B)$ is a uniform distribution, then nothing is known about the B-values. Following Bohr, observables A and B, which are in mutually exclusive relation such that a knowledge of A makes any prediction of B impossible, and vice versa, are designated as *complementary*. The best-known example of two complementary quantities is the position x and the momentum p of a system.

With the aid of this quantum-mechanical theory, it now becomes possible to trace the measuring process in detail by explicitly taking into consideration the interaction between the measuring device and the system observed. The emphasis must rest on the question why in the measurement of a property that is not objective, and that is associated with a change of state, uncertainties appear that enter as probabilities into the framework of quantum theory. It is not yet apparent how the fact that during the measuring process essential parts of the originally available information on the system can be lost, is to be expressed in a closed mathematical theory.

Although it is restricted in comparison to classical physics, the possibility of experimental knowledge of a quantum-physical system, and especially the concept of state, received explicit consideration in the construction of quantum theory. The basis for this restriction of the possibilities of knowledge is to be found in the physical laws of the measuring devices. But these laws are part of just this quantum theory. Quantum theory will therefore be consistent with its assumptions only if these assumptions, which determine the possibilities of experimental experiences, can themselves be derived from a theory of the measuring process within the framework of the quantum-mechanical formalism. As in the theory of relativity, this novel structure of the theory can in turn be traced to the fact that experimental investigations can only be carried out with measuring instruments that themselves are governed by the laws of physics, in this case quantum theory. The central significance of a theory of the measuring process is evident from this structure.

3. THE MEASURING PROCESS

For the exposition of the measuring process within the framework of quantum theory[1] we start with a system S and a measuring device M. The process for measurement of an observable A with the measurement value A_i consists in the following: In a certain time interval Δt an interaction occurs between the system S and the measuring device M; subsequently the value A_i of the observable A is determined from the measuring device.

Suppose that the system S is in state $|\varphi\rangle$, and the measuring device in state $|\Phi\rangle$. Let the Hamiltonian of $S + M$ be decomposed into the components $H = H_0 + H_w$. H_w describes the interaction that occurs between S and M in the time interval $t = 0$ to $t = t'$. Then the time dependence of the states $|\varphi\rangle$ and $|\Phi\rangle$ can be ascribed to H_w alone. We thus use an interaction representation with respect to H_w.

The measuring process then occurs in several steps.

3.1. *Decomposition and Extension*

The state $|\varphi\rangle$ and the operator $P_\varphi = |\varphi\rangle\langle\varphi|$ are expanded in terms of the spectrum $|A_i\rangle$ of the operator A:

$$|\varphi\rangle = \sum_i \langle A_i|\varphi\rangle|A_i\rangle$$

$$P_\varphi = \sum_{i,\,k} |A_i\rangle\langle A_i|\varphi\rangle\langle\varphi|A_k\rangle\langle A_k|.$$

where $|\langle A_i|\varphi\rangle|^2 = w_\varphi(A_i)$ is the probability that after the measuring process the eigenvalue A_i will be found for A. The next step consists in a formal extension of the Hilbert space by including the measuring device in our considerations. For this purpose we introduce the new states:

$$|\Psi\rangle = |\varphi\rangle|\Phi\rangle \qquad |\Psi_i\rangle = |A_i\rangle|\Phi\rangle$$

From the above expansion one obtains for the state and the projection operator of the systems $S + M$:

$$|\Psi\rangle = \sum_i \langle A_i|\varphi\rangle|\Psi_i\rangle$$

$$P_\Psi = |\Psi\rangle\langle\Psi| = \sum_{i,\,k} |\Psi_i\rangle\langle\Psi_k|c_{ik}$$

$$c_{ik} = \langle A_i|\varphi\rangle\langle\varphi|A_k\rangle.$$

The expansion as well as the extension represent purely formal steps that can also be undertaken in the inverse order. True physical events do not occur until the second part of the measuring process, reception.

3.2a. *Reception*

Suppose the interaction H_w between S and M begins at time $t = 0$ and continues to time $t = t'$. During this time interval the state $|\Psi\rangle$ undergoes a change. If

$$|\Psi(t = 0)\rangle = |\Psi\rangle,$$

then, for $0 \leqslant t \leqslant t'$

$$|\Psi(t)\rangle = e^{iH_wt}|\Psi\rangle$$

and

$$|\Psi_i(t)\rangle = e^{iH_wt}|\Psi_i\rangle.$$

After the interaction the joint system $S + M$ is in the state

$$|\Psi'\rangle = |\Psi(t')\rangle = \sum_i \langle A_i|\varphi\rangle|\Psi_i'\rangle,$$

where

$$|\Psi_i'\rangle = |\Psi_i(t')\rangle.$$

which for $t > t'$ is again constant. Furthermore

$$P_{\Psi'} = |\Psi'\rangle\langle\Psi'| = \sum_{i,\,k} |\Psi_i'\rangle\langle\Psi_k'|c_{ik}.$$

The following difficulty now arises: The state $|\Psi'\rangle$ can no longer be decomposed to a product of two states, one of which describes the object and the other describes the measuring device. The states

$$|\Psi_k'\rangle = e^{iH_wt'}|A_k\rangle|\Phi\rangle = |A_k'\rangle|\Phi_{A_k'}\rangle$$

continue to be separable, provided the measuring device is really suitable for the measurement of the quantity A, but with states $|\Phi_{A_k'}\rangle$ which depend on the observed value A_k' of the object. Thus the joint state is given by

$$|\Psi'\rangle = \sum_k \langle A_k|\varphi\rangle|A_k'\rangle|\Phi_{A_k}'\rangle$$

and is no longer separable. For a measuring device that introduces no distortion, and hence is of any desired sensitivity, we have $|A'_k> = |A_k>$

$$P_{\Psi'} = \sum_{i,k} |\Phi_{A_i}>|A_i><A_k|<\Phi_{A_k}|c_{ik}.$$

If H_w is Hermitian, then the states $|\Phi_{A_k}>$ of the measuring device are normalized. They are not, in general, orthogonal. However, M can be considered to be a measuring device that is suitable for the measurement of A, only if M is constituted so that the $|\Phi_{A_k}>$ are orthogonal, i.e. if

$$<\Phi_{A_k}|\Phi_{A_i}> = \delta_{ik}$$

is valid. Only then is it possible to unambiguously infer the value A_k of the object from a knowledge of the state $|\Phi_{A_k}>$ of the measuring device. The new states $|\Phi_{A_k}>$ must be orthogonal to the initial state $|\Phi>$ as well, otherwise they cannot clearly indicate the measurement result.

3.2b. *Restoration*

The orthogonality of the states $|\Phi_{A_k}>$ of the measuring device is also important for another reason. To demonstrate this, we consider a Stern-Gerlach experiment in which a beam of atoms in state $|\varphi>$ is split by an inhomogeneous magnetic field F_1 into two spatially separated components that correspond to the two possible spin orientations A_1 and A_2. We again denote the states of the two spin components by $|A_1>$ and $|A_2>$.

Suppose that after a short distance the split beam of atoms is brought to interference by a second field F_2. This action causes no difficulty at this point.

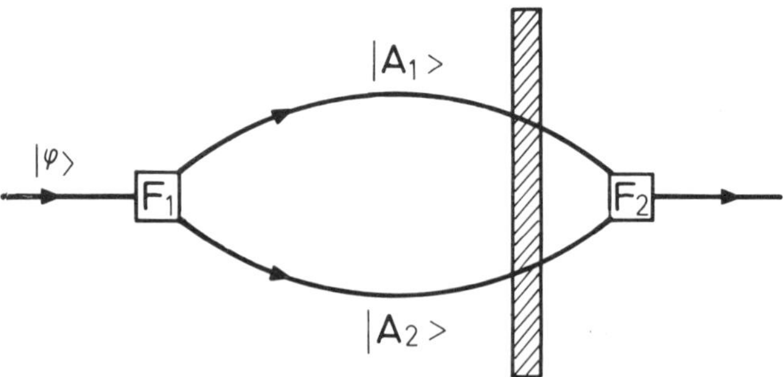

Fig. 8. Stern-Gerlach experiment with restoration.

The situation is changed entirely, however, if a measurement of position is introduced between the two magnetic fields F_1 and F_2, in order to ascertain the spin of the respective particles. We do not yet wish to consider the measurement of position itself, but rather the necessary interaction of the particles with the measuring device, a photographic plate, for example. Once this has occurred, then the back interference of the two beams is possible only if one succeeds in separating the measuring device from the object by restoration. For this to occur, it is necessary that the states $|\Phi_{A_k}\rangle$ again pass over to a certain state $|\Phi''\rangle$, the initial state $|\Phi\rangle$, for example. In that case the state

$$|\Psi'\rangle = \sum_k \langle A_k|\varphi\rangle|A_k\rangle|\Phi_{A_k}\rangle$$

separates into a product

$$|\Psi''\rangle = |\Phi''\rangle \sum_k \langle A_k|\varphi\rangle|A_k\rangle,$$

and the states $|A_k\rangle$ can once again interfere. But since measuring devices, as for example photographic plates, in general have macroscopic dimensions, and hence have approximately 10^{20} degrees of freedom, the restoration of a unified state $|\Phi''\rangle$ of the measuring device will cause considerable difficulties. Even a partial back interference is hardly attainable. It is important to note, however, that even after the interaction with the measuring device, a back interference of the object states is possible at least in principle. At this stage of the measuring process it is consequently objectively still undecided which value A_i can actually be ascribed to the object system.

To bring about a decision, those components of the projection operator $P_{\Psi'}$ that make a back interference still possible must be caused to vanish. These are the mixed terms. Furthermore, the object S must be separated from the measuring device, so that a proposition concerning the value A_i of the observable A in the object system can be obtained.

3.3. *The Cut (or Separation)*

Isolating the object system from the measuring apparatus is the most important part of the measuring process. By this procedure, designated the cut, the interference terms, which made an objectivation of the property A impossible, vanish.

In order to obtain from

$$P_{\Psi'} = \sum_{i,\,k} c_{ik} |\Psi'_i\rangle\langle\Psi'_k|$$

a proposition about the object alone, it is necessary to revoke the extension of coordinate space. This is achieved by projection on that part of Hilbert space that corresponds to the object S. By this operation of 'reduction' the operator $P_{\Psi'}$ becomes the statistical operator

$$W_\varphi = \sum_i c_{ii} |A_i\rangle\langle A_i|$$

of the object system, where $w_\varphi(A_i) = c_{ii}$ is the probability of finding the value A_i. W_φ represents a mixture of the object system alone. Abstracting the measuring device thus has the consequence that the property A becomes an objective property; the theory does not, however, declare anything about the particular value of A, but only gives a probability distribution $w_\varphi(A_i)$.

It is conspicuous that the theory of the measuring process makes no assertion as to the exact value of A, but only provides probabilities. A loss of information is in fact caused by the cut. After the reception, there existed a strong correlation between the states of the object and the states of the measuring device. These were described by the information contained in $P_{\Psi'}$. By the mental separation of the measuring device from the object, all information that relates to the correlation is lost; only the propositions contained in $P_{\Psi'}$ that relate solely to the object or to the measuring device are retained.

The loss of information caused by the cut can at once be quantitatively calculated if we introduce as a measure of the subjective unknown the entropy

$$S = -k\,Sp\,(W \lg W),$$

that is associated with the statistical operator W. Prior to the cut

$$S_{\Psi'} = -k\,Sp\,(P_{\Psi'} \lg P_{\Psi'}) = 0$$

and after the cut

$$S_\varphi = -k\,Sp\,(W_\varphi \lg W_\varphi) \geqslant 0.$$

Because of this loss of information it is necessary to make probability assertions about the system S. The situation in which the description of a system contains subjective elements (in this case a lack of knowledge) is not new in physics. Thus, in statistical mechanics physical systems are characterized by a probability distribution in phase space.

The transition from $P_{\Psi'}$ to W_φ can also be accomplished by dispensing with the separation of object from measuring device and bringing the interference terms to vanish in the joint system $S + M$. By this step those observables of the joint system, whose eigenstates are characterized by $|A_k\rangle|\Phi_{A_k}\rangle = |\Psi'_k\rangle$, become objective properties. For this reason we call this part of the measuring process 'objectivation'. By objectivation, a mixture with the statistical operator

$$W_{\Psi'} = \sum_i c_{ii}|\Psi'_i\rangle\langle\Psi'_i|.$$

results from $P_{\Psi'}$. This operator still refers to the joint system $S + M$. To obtain a proposition solely about S, the measuring apparatus must still be abstracted. After the objectivation has been carried out, however, this abstraction is a physically insignificant, purely formal step. By reduction the statistical operator W_φ results from $W_{\Psi'}$, and now relates to the object system only. The reduction of $W_{\Psi'}$ is physically insignificant inasmuch as the statistical uncertainties contained in W_φ were already present in $W_{\Psi'}$, so that no loss of information is entailed by the abstraction from the measuring instrument. The entropies $S_{W_{\Psi'}}$ and S_{W_φ} of the statistical operators $W_{\Psi'}$ and W_φ, respectively, are equal:

$$S_{W_{\Psi'}} = S_{W_\varphi}.$$

3.4. *Registration*

As in statistical mechanics, the last step in the quantum-mechanical measuring process, which finally yields an exact proposition about the object, occurs simply in that the scale of the respective measuring instrument is read, and it is determined which state $|A_i\rangle$ is actually realized. By this step, which we call 'registration', the entropy S is once more reduced to zero. The reason for this is that the information which the observer has about the object, is enlarged by this act of knowing.

Registration concludes the measuring process. It is an unproblematic part of this process, insofar as no specifically quantum-mechanical processes occur, but only those that are already known from classical statistical mechanics.

We schematically summarize the different steps of the measuring process (Figure 9). The process starts at P_φ and ends at P_{A_i}. The arrows indicate the sequence of the different possible steps. On the left side are the states of the object system S, on the right those of $S + M$.

In summary, the following can be said: The measurement of a non-objective property A for a system in the state $|\varphi\rangle$ proceeds so that after a

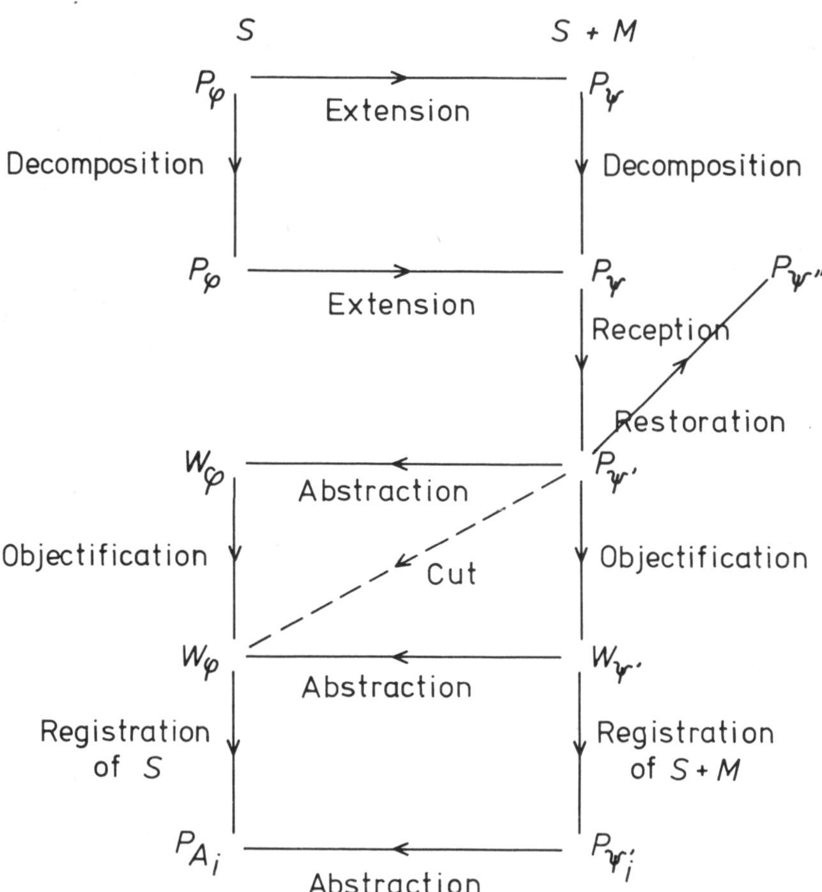

Fig. 9. Schematic representation of the measuring process.

formal extension of the coordinate space $P_\varphi \to P_\Psi$ the system S is brought into a real interaction with the measuring instrument $(P_\Psi \to P_{\Psi'})$. This causes the state of the system to transform to a different state $|A_i\rangle$, to which the property A can be objectively attributed. In order to establish this, the object must be separated from the measuring device, which is accomplished by the cut $(P_{\Psi'} \to W_\varphi)$. Associated with the cut is a loss of information. For this reason the actual value of A can only be predicted as a probability. Registration then establishes which value A_i actually occurs $(W_\varphi \to P_{A_i})$. The associated gain in information then again returns the entropy to its original value, $S = 0$.

In this presentation of the measuring process, the formal mathematical treatment of the problem was given prominence; however, several questions of great physical importance remained unanswered. These questions concern the cut, or separations. For it is still entirely unresolved just why the interference of $P_{\Psi'}$ can in fact be caused to vanish. Closely connected with this is the problem how, ultimately, the subjective element associated with the cut is introduced into the theory, as indicated by the appearance of probabilities. The loss of information during the objectivation, represented by an increase in entropy, as well as the subsequent increase of knowledge during registration indicate that in the quantum-mechanical measuring process the observer, or rather the knowledge of the object by the observer, has attained a new significance that was completely unknown in classical physics.

4. THE CUT

It is the function of the measuring process to state which value A_i of an observable A can be ascribed to the system S, if prior to the measurement the system was in state $|\varphi\rangle$. Since the system S interacted reciprocally with the measuring device M during a time interval Δt, it remains associated with the measuring device in a very complicated manner. It is necessary therefore to isolate the object from the measuring device once again. This separation we have called the cut (or simply 'separation'). It was further demonstrated that this cut is necessarily associated with a loss of information. The state of the system transforms to a mixture, so that only probability assertions can be made about the measurement values of the object.

These difficulties suggest, of course, that the uncertainty can be avoided if the cut between the object system S and the measuring device is not at all performed. Then the task would consist in making a measurement with respect to the property A for the joint system $S + M$, i.e. to determine which of the states $|A_k\rangle|\Phi_{A_k}\rangle = |\Psi'_k\rangle$ is present. But this would only be possible with a second measuring device M'. For this second instrument the preceding considerations regarding its mode of operation would apply equally, however. If one starts with $P_{\Psi'}$, then a proposition on the state $|\Psi'_k\rangle$ of $(S + M)$ can only be obtained by performing a cut between $S + M$ and M', which then gives the mixture $W_{\Psi'} = \Sigma|\Psi'_k\rangle\langle\Psi'_k|c_{kk}$. This is the step that, in the schematic representation of the measuring process, corresponds to the transition $P_{\Psi'} \rightarrow W_{\Psi'}$, and that we called objectivation. As previously mentioned, the mixture $W_{\Psi'}$ contains the same uncertainties as does the operator W_φ that relates to the system S.

For the result of a measurement it is therefore irrelevant whether one passes from $P_{\Psi'}$ to W_φ directly by abstraction from the measuring instrument and thereby generates a mixture, or whether, to begin with, one observes $S + M$ together using a second measuring device M'. It will eventually be necessary to abstract from M' as well if one intends to obtain information on $S + M$, and in this abstraction the same uncertainties appear as before. The fact that in some respects it is immaterial whether the cut is inserted between S and M or between $S + M$ and M', is called the displaceability of the cut. For the measuring process, it is merely essential that the cut somewhere between the object and the perceiving subject be made at some time.

At which position it occurs is unimportant for the outcome of the measurement.

So far these considerations have not shown how the cut actually takes place, i.e. why in the course of the measurement information is lost, so that the measuring process becomes irreversible. Nothing is decided about the actual value of A by the interaction with the measuring device. The information originally present is conserved during the entire interaction and is still completely available at the end of the process.

The question therefore arises how, subsequent to the reception, the objectivation of the states $|\Phi_{A_i}\rangle|A_i\rangle$ and the cut are actually to take place. We briefly discuss various explanations that have been suggested:

(a) The interpretation of the cut as a subjective process: Which of the states $|\Psi_i'\rangle$ has been realized after the reception can be decided only by involving the observer as cognitive subject in the process. Eliminating the interference from the part of the measurement that we called objectivation only makes sense if a human observer establishes that, for example, the state $|A_k\rangle|\Phi_{A_k}\rangle$ has been realized. It is then reasonable to eliminate those cases not realized and, of course, the interference terms as well. The transition to the object state $|A_k\rangle$ is then entirely unproblematical.

Against this interpretation the objection can be raised that a function is ascribed to the cognitive subject, or to the act of cognition, that is entirely outside what is physically possible. For as long as $|\Psi'\rangle$ is present, and not a mixture $W_{\Psi'}$, it is objectively undecided which state $|\Psi_k'\rangle$ has been realized. The assumption that a certain $|\Psi_k'\rangle$ is present, but that the observer can have no knowledge of it, immediately leads to logical contradictions, as will be shown in our following chapter.

Beyond this, an observer can only take cognizance of what is already established. For the act of cognition must physically consist of an interaction between $(S + M)$ and the brain of the observer. If the interference terms have not vanished by the interaction of S and M, then they will also not vanish by an interaction with the brain. Thus even in the brain a superposition, but not a mixture, will be present; the observer will therefore not experience the subjective sensation of having perceived something definite. As, for example, in the observation of a Young's interference experiment with a photon, he will on the contrary not be able to decide whether the photon passed through one or the other slit, since his view on this will change continuously.[2]

The act of cognition can therefore only relate to a situation in which certain facts are objectively determined; these, however, are known to the observer only in part, in the form of probabilities, for example. Such a situation is a mixture, however, as in this case $W_{\Psi'}$. If, on the other hand, the interference terms are still present, as in $P_{\Psi'}$, then there exists no real possibility of performing the registration. The interference terms will in fact prevent the registration, since cognition of something can only occur if it is established.

It thus appears that the act of cognition is not suitable for eliminating the interference terms. But since it is in fact possible to make measurements, there must consequently exist another mechanism that accomplishes the transformation to a mixture. The interpretation of the cut as an objective process is an attempt to specify such a mechanism.

(*b*) This interpretation starts with the observation that measuring devices, in general, are of macroscopic dimensions, i.e. that they have approximately 10^{20} different degrees of freedom. Even if the measuring device was in state $|\Phi\rangle$ at the beginning of the measurement, almost all degrees of freedom become important for the overall state during the interaction with the object system. The probability that this state of the system, which is now in a very high dimensional coordinate space, again reverts without special intervention to the simple initial state decreases, the larger the number of degrees of freedom. Also correspondingly small is the probability for the possibility of a back interference of the object states, since this is only possible for a complete restoration of the measuring instrument. If the measuring instrument is sufficiently large, i.e. if it has macroscopic dimensions, then the interference terms will become arbitrarily small and are practically never of any consequence. The possibility of a back interference is therefore not destroyed by the cognition of the observer, but rather by the interaction between the object and the measuring device. The measuring process is therefore to be regarded as irreversible on the basis of the physical events. The result of the measurement is determined, even before it has been observed. Cognition by the observer has no effect on the outcome of the measuring process.[3]

This situation can be illustrated by a mental experiment attributed to Schrödinger. The following items are placed in a closed chamber: a radioactive atom; a Geiger counter, which on responding, activates a lever that breaks a glass container of cyanide; and a cat. The question whether several hours after the assembly of the experimental set-up the cat is dead or alive cannot be resolved by quantum theory. The actual state of the cat will correspond to a superposition of the state of death and the state of being alive. In the sense of the subjective interpretation of the cut, only an observation can decide whether the cat is dead or alive.

The objective interpretation would point out that for this example the application of quantum-mechanical concepts does not make sense. The cat, as a macroscopic object that is used for the measurement of the radioactive decay, is in a mixture, well before the observer opens the chamber. The state of death of the cat is a terminal state of a macroscopic irreversible process that can no longer interfere with being alive. The irreversibility of the part of the measuring process, which alone can be explained by objective-physical processes, is particularly apparent in this example.

But in spite of this suggestive example, this interpretation must be challenged by the objection that no physical process, even when it appears to be as irreversible as the dying of the cat, can make the interference term exactly zero. For in principle the entire information that the observer had at the beginning of the measuring process is conserved. An increase in entropy does not take place nor does a transition of a state to a mixture.

The difficulties, which appear to make it impossible to interpret the cut as either a subjective or an objective process, lead to a further attempt at explanation. In this approach the cut is interpreted as a mental process.

(c) In this interpretation the interference terms are caused to vanish neither by purely subjective nor by purely objective processes. It should be recalled that the original reason for the cut was to make possible a proposition about the object without further regard for the measuring device. To begin with, the abstraction from the measuring instrument is purely a mental process: To accomplish a separation of the object from the measuring device and thereby to facilitate a proposition about the object alone, the observer dispenses with the information contained in the interference terms. The consequence of this voluntary sacrifice is a loss of information, which finds expression in the appearance of probabilities, i.e. by the cut the system $P_{\Psi'}$ undergoes a transition to a mixture $W_{\Psi'}$.

This interpretation is very close to the mathematical description of the cut process. In fact, by the operation of reduction, precisely those components of the information contained in $P_{\Psi'}$ are projected that relate solely to the object. On the other hand, the part of the information that relates to the correlation between object and measuring device is destroyed by the operation of reduction, since it cannot be used for the characterization of the object alone.

It is also correct that the observer voluntarily relinquishes certain parts of the information in order to isolate the object system by this mental process. But a sacrifice of knowledge about the interference ability of the object is reasonable only if the object is in fact no longer capable of participating in an interference process. Otherwise one would go from a complete to an incomplete description of the situation, without any justification from the circumstances to do so. It only makes sense to neglect information if it no longer has any real significance.

It thus appears that the purely mental act of abstracting from the measuring instrument is also not able to appropriately interpret the process of the cut. The three attempts at interpretation discussed here in fact only select respective partial aspects of the total phenomenon. We shall see that an appropriate interpretation must give consideration to the entire very complex situation.

Although the separation of the object system from the measuring device is purely a mental process in which information is lost, this process can be regarded as reasonable only under certain real conditions. One of these conditions is that a measuring device of macroscopic dimensions be used. In an interaction of this device with the object the information, initially present in compact form, distributes itself over approximately 10^{20} degrees of freedom of the measuring device and thereby – at least to a certain extent – becomes practically inaccessible. For even the purely computational pursuit of the complicated system of equations, which are given in the high dimensional coordinate space of $S + M$, is entirely beyond any actual means. Thus the situation after the reception is that essential parts of the information must be regarded as practically lost because of the extension of the coordinate space. A realizable possibility of using this information further does not exist, either experimentally or theoretically.

The abstraction from the measuring device consists in mental acknowledgement of this situation. By voluntarily abandoning those parts of the information that are for all intents and purposes completely inaccessible, the observer isolates the object from the measuring device. The cut thus loses all forceful features. It is no more than the acceptance of a situation which already exists in practice. But it thereby becomes possible to advance propositions about the object alone without reference to the measuring device.

From these considerations it follows that a sacrifice of information only makes sense if measuring devices are used that have sufficient degrees of freedom, thus ascertaining that the information lost can practically never be of any importance. Only then will the theoretical propositions, made on the basis of the cut, be in harmony with the experiment. This confirms the condition established by Bohr that physical measuring devices must always have macroscopic dimensions.

It is the goal of the quantum-mechanical measuring process to obtain propositions about the values of a non-objective property A of a system S. This can only be accomplished if the state $|\varphi>$ of S undergoes a transition to the eigenstate $|A_i>$ of A by an interaction with the measuring device. The interaction, however, has the consequence that the measuring device is connected with the object in a manner that is in general irreversible. A separation of the object from the measuring device by a mental cut is only reasonable if the measuring device has so many degrees of freedom that essential parts of the information initially present are practically lost in the course of the interaction process. The section is then no more than the neglect of these parts of the information, which would practically have been lost anyway. An entropy increase is associated with this, which cannot be derived from the quantum-theoretical equations.

The appearance of mixtures and the associated uncertainty of the measuring results is thus a necessary consequence of the fact that the object must be separated from the measuring device. It is important to note that the quantum-mechanical uncertainties are not produced by an acausal mechanism unknown to us, but arise exclusively from the sacrifice by the observer of certain parts of the information initially present. The methodological requirement of physics, to describe objects, thus imposes an uncertainty in the prediction of the properties of these objects.

5. THE FUNCTION OF THE OBSERVER IN QUANTUM THEORY

The preceding considerations of the measuring process make it possible to answer the question, to what extent the quantum-mechanical description of reality assigns to the observer a special function and thereby introduces subjective elements into the theory, which were still unknown in classical physics.

The description of a physical system by its state $|\varphi\rangle$ is the most comprehensive description of a system that is possible in the framework of quantum theory. But at first it would appear that this characterization is not an objective description of reality, for the expansion of $|\varphi\rangle$ in terms of the states $|A_i\rangle$

$$|\varphi\rangle = \sum_i |A_i\rangle\langle A_i|\varphi\rangle$$

contains coefficients $\langle A_i|\varphi\rangle$, which we previously associated with the probabilities $w_\varphi(A_i) = |\langle A_i|\varphi\rangle|^2$ for the occurrence of the value A_i. Probabilities, however, are prepositions about knowledge of the object by the subject. Thus it could be presumed that $|\varphi\rangle$ is not an objective description of the system S, but rather that $|\varphi\rangle$ represents a proposition about knowledge of the object by the observer.

In this form however, the assertion is not quite correct. The fact that $|\varphi\rangle$ is the state of the system is an objective property of S in the sense that the operator $P_\varphi = |\varphi\rangle\langle\varphi|$ has the value 1 and not 0. The coefficients $\langle A_i|\varphi\rangle = \varphi(A_i)$ are nothing other than a special representation of the state $|\varphi\rangle$; their ensemble only state that the system S just finds itself in this state $|\varphi\rangle$. $|\varphi\rangle$ or P_φ is therefore to be understood as an objective description of the system S.

With regard to possible measurement of a non-objective quantity A with the values A_i, one obtains from $|\varphi\rangle$, on the other hand, probability statements $w_\varphi(A_i) = |\langle A_i|\varphi\rangle|^2$, i.e. statements that explicitly concern the knowledge of the observer. These probabilities do not relate to the knowledge of the object in state $|\varphi\rangle$, but to the knowledge of the observer, prior to the registration, about how the object system will be constituted after a measuring process on the quantity A.

The probability statements thus do not at all refer to the pure case P_φ, but to the mixture W_φ. The statistical operator W_φ, however, is a statement about the observer's knowledge about the object, insofar as information is lost in the transition $P_{\psi'} \to W_{\varphi}$, i.e. the cut. This loss of information caused by the cut is the reason why, in quantum theory, probabilities, i.e. statements about the knowledge of the observer, occur. The subjective element that enters quantum-mechanical theory in this way is essentially not different from the subjectivity familiar from statistical mechanics. The only, but essential, difference is that in classical physics, statistical propositions can be avoided, at least in principle. In quantum theory the difficulties that are caused by the separation of the object from the measuring device do not allow the statistical element to be avoided in principle.

But as long as a measurement of the quantity A has not yet been performed and the system is still in state $|\varphi\rangle$, the description of S does not contain any subjective features. The coefficients $\langle A_i|\varphi\rangle$ only indicate precisely the occurrence of $|\varphi\rangle$. The interpretation of the quantities $|\langle A_i|\varphi\rangle|^2$ as probabilities for the occurrence of A_i is not yet possible at this stage. For already the hypothetical assumption (that one can assign the non-objective observable A to the system in state $|\varphi\rangle$ in a manner in which one of the values A_i occurs but for which the observer can only make probability statements $|\langle A_i|\varphi\rangle|^2$), leads to logical contradictions, as will be shown in the following chapter. Of a particular property E that corresponds to the operator $P_{A_i} = |A_i\rangle\langle A_i|$ one can, in the state $|\varphi\rangle$, assert neither that it occurs, nor that it does not occur, since both assertions lead to contradictions. In the following chapter the investigation of the problem of substance, especially the question of the possibility of objectification of properties, will make this clear.

NOTES

[1] Cf. e.g. G. Süssmann, *Uber den Messvorgang*, Munich 1958; J. M. Jauch, *Foundations of Quantum Mechanics*, Reading, Mass. 1968, p. 183.
[2] As the following considerations will show, the interference of impressions can never occur in practice. Because of their physical properties, the sense organs and the brain of man make a back interference practically impossible. But in principle this cannot be excluded.
[3] It is by no means necessary to use photographic plates or other recording devices for the elimination of the interferences. The sense organs and the brain of the observer are already of the order of magnitude that renders a back interference virtually impossible. This explains the previous comment that an interference of sense impressions can practically never occur.

CHAPTER IV

.

THE CONCEPT OF SUBSTANCE

The physical conditions for knowledge of atomic systems have the consequence that not all measurable properties can be attributed in the same way to an observed object. The philosophically relevant formulation of this circumstance is related to a critique of the traditional concepts of substance: Under certain circumstances, constraints, determined by the possibilities of cognition, are imposed on the applicability of the category of substance to physical events. A discussion of our concept of objectifiability explicitly demonstrates the impossibility of relating all measurable observables to a well-defined object.

1. THE CONCEPT OF SUBSTANCE IN CLASSICAL PHYSICS

The concept of substance originated in the philosophy of Aristotle. Descartes was first to discuss it in the form in which it became significant for modern natural science. In posing the question of what a piece of 'wax' truly is, when it changes its properties[1], Descartes arrives at the conclusion that 'wax' is different from all that is perceived by the senses and that it is by the mind alone that it can be apprehended. Whereas the perceptible properties of the wax, the accidents are changeable, the wax as such is unchanged but also inaccessible to direct perception.

Attributing accidental properties to a substance is therefore an intellectual process. Wax is that which persists in the course of change of the apparent form; it is that to which we refer the changing properties, which we thus objectify.

It is already apparent at this point that the concept of substance is essentially a point of view for ordering and subsuming phenomena. With Hume the suspicion first appears that the speaking of things in the form of attributing a predicate P to a subject S is the motive for ascribing an inherent existence to the S to which all these predicates relate. Hume therefore regards the concept of substance to be a misconception, because nothing in our experience corresponds to it:

... our propension to confound identity with relation is so great, that we are apt to imagine something unknown and mysterious, connecting the parts, beside their relation; ... which view of things ..., obliges the imagination to feign an unknown something, or *original* substance and matter, as a principle of union or cohesion among these qualities, ...[2]

Against this thesis of Hume the objection is raised by Kant that substance is a category (the relation between inherence and subsistence), and as such is one of the conditions by which things at all become manifest in experience. It is only by the interpretation of phenomena as accidents of a substance, variable in time, that things come to be in our experience. In view of this experience-constituting function of the category of substance, the identification of a substance enduring through the changes of appearance should always be possible, at least in principle. It should therefore be possible consistently to relate all properties of a thing appearing in experience to a substance that endures in spite of the transitoriness of the accidents. This finds expression in Kant's principle of persistence of substance – a synthetic, but *a priori* valid, proposition: "All appearances contain the permanent (substance) as the object itself, and the transitory as its mere determination, that is, as a way in which the object exists."[3] Whereas Hume's empiricism asserts that one can always relate all possible properties to a substance ("an unknown something"[4]), but that it is meaningless to do so, Kant is more cautious by asserting this possibility only for objects of experience, but doing so *a priori*. Nowhere does Kant state that all phenomena whatever can be ordered into object experiences; but he does assert that whenever this is not possible no cognition of any kind can occur.

The justification of the principle of persistence of substance must derive from the conditions for the possibility of experience, in this case from the role of the category of substance in subsuming phenomena under objects of experience. This is related to the general question, "how pure concepts of the understanding can be applicable to appearances".[5] The mediating idea, akin to both category and phenomena, Kant designates as the transcendental schema. He argues further that the schema of pure concepts of the understanding is the transcendental determination of time. For the categories of relation (substance, causality, reciprocation) this means that the schemata of these categories are the time determinations that refer to the temporal order of phenomena. Regarding substance, in particular, it is then said that: "The schema of substance is permanence of the real in time."[6] This reference to time is expressed even more strongly in Kant's second edition in the formulation of the law of conservation of substance: "In all change of appearances substance is permanent; its quantum in nature is neither increased nor diminished."[7] This law is *a priori* valid because the existence of things in our experience already presupposes substance that endures through the changes of appearances: "Permanence of substance is thus a necessary condition under which alone appearances are determinable as things or objects in a possible experience."[8] "This permanence (of substance) is, however, simply the mode in which we represent to ourselves the existence of things (in the field of) appearance."[9]

Thus whenever it is possible to consolidate phenomena as experiences of things by relating them as possible accidents of a substance that is immutable through the change of phenomena, the law of the persistence of substance is valid. Conversely, however, the fortuitous immutability of a property, e.g. a conserved quantity, is by no means sufficient for interpreting it as the substance of the object. For the concepts of temporal change and persistence cannot be defined without reference to substance. Persistence, as Kant noted, is the way substance appears in the phenomenon. "To time, itself non-transitory and abiding, there corresponds in the field of appearance what is non-transitory in its existence, that is, substance. Only in relation to substance can the succession and coexistence of appearances be determined in time."[10] The *a priori* validity of the law of conservation of substance then follows from the association of the concept of substance with the concept of time. An arbitrary property that does not change with time, as a consequence of physical laws, is by no means to be regarded as substance, since the immutability of a phenomenon merely indicates that it experiences no change with respect to the substance. This finds expression in that the laws of conservation which are important for a physical process in general cannot be regarded as *a priori* valid.

The constancy of a certain property therefore is by no means sufficient reason for inferring the constancy of the bearer of this property. In this connection Kant states: "All that we know of matter is merely relations . . . but among these relations some are self-subsistent and permanent, and through these we are given a determinate object."[11] The conserved quantities provide the possibility of characterizing a certain object as such by its properties, but not necessarily that of a substantiation of that object. It is decisive for the conception of substance that the possibility exist to consistently relate any arbitrary property of a thing to a substance. Only from this function of the category of substance can the principle of the persistence of substance be explained. But, as has been mentioned, it may not be interpreted as a conservation law in the physical sense. Only by the persistence of substance is it possible to explain what is to be understood either by a temporal change or by persistence.[12]

This concept of substance can be applied to all things in our experience, since objects of experience can only come to exist by application of the category of substance to the manifold of perceptions. As Kant notes, 'things' are always subject to the principle of 'complete determination'[13], i.e. for every property, either the property or its converse is applicable to them. Since, as we know, the objects of classical physics are always 'completely determinable', i.e. the measurement of any arbitrary property can be performed on them, there is no objection to the application of the concept of substance to objects of classical physics. The ordering of phenomena within classical physics can therefore always be accomplished with the aid of this category.

We will show, however, that the applicability of this concept of substance within atomic physics is restricted by certain bounds, which are determined by the physical conditions of the possibility of experience, i.e. by the laws of the measuring instruments.

2. THE CONCEPT OF SUBSTANCE IN QUANTUM THEORY

Reflection on the physical conditions by which alone experimental results can be gained has shown that for a quantum-theoretical system only commensurable properties can be measured without affecting the system by this measurement. The ensemble of the properties measured in this way is symbolically represented by the state vector $|\varphi\rangle$ and by the projection operator P_φ that projects on $|\varphi\rangle$.

If, to begin with, one neglects the possibility of measuring non-objective properties by a measuring process or of making a probability statement about its outcome, then only the objective properties are available for direct observation. Application of the category of substance to these properties, i.e. consolidation of the respective experiences into objects, poses no difficulties. The ensemble of commensurable properties can always be referred to a distinct object system, which can then be represented by the corresponding state vector (in the Heisenberg representation). In view of the concepts used, the circumstances in quantum theory are even more lucid than in classical physics. The theory introduces by means of the state $|\varphi\rangle$ a specific symbol for substance, whereas in classical mechanics the object, to which observables refer, is usually identified only by an index appended to this quantity.

The 'quantum-theoretical concept of substance' defined by the state $|\varphi\rangle$ has nevertheless essential features of the traditional concept of substance. The state $|\varphi\rangle$ persists in time, whereas the observables, as accidents of the substance, change with time. Each objective property can without difficulty be referred to the state $|\varphi\rangle$ in the sense that each property can either be attributed to the system identified by $|\varphi\rangle$, or not. With this quantum-mechanical concept of substance, the difference between the substance and any arbitrary conserved quantity can be very distinctly clarified: The fact that $|\varphi\rangle$ or P_φ do not vary with time is not to be regarded as a conservation law; rather it is the constancy of $|\varphi\rangle$ that defines what is meant by a change with time.[14]

Only at one point does the concept of substance, as used here, deviate from that of classical physics: It is not the carrier of all measurable properties, but only of those (objective) properties, which can be measured on the system without changing the state $|\varphi\rangle$. The concept of substance restricted in this manner does not conform with the concept of thing used in classical physics and in the philosophical tradition. Thus Kant, for example, states:

> Every thing . . . is likewise subject to the principle of complete determination, according to which if all the possible predicates of things be taken together with their contrary opposites, then one of each pair of contradictory opposites must belong to it.[15]

But it is precisely this that is not the case for the 'things' of quantum mechanics. It is not apparent, however, to what extent the restriction of the concept of substance, which became important with the limitations on the means of cognition, could be detrimental. The function of the category of substance by which experience is at all constituted, is no more restricted than the *a priori* validity of the principle of the persistence of substance derived from this function.

On the other hand, fundamental difficulties result if one attempts to extend this concept of substance, derived from the possibility of cognition, in the sense that one regards it as a carrier of all measurable properties, as in classical physics. It will become apparent that it is precisely the function of relating different properties to an object, thus objectifying them, that is lost; but this is a function of decisive importance to the category of substance.

There are various reasons for introducing a figurative quantum-mechanical concept of substance in this manner, one that is not literal. Suppose the observed system is an elementary particle, a proton for example. Then, the proton persists as a proton through all changes with time that the properties undergo according to the Schrödinger equation, even when these are not commensurable. Its rest mass, charge and spin, and thus all those properties that identify the proton as a proton, remain unchanged. Even a change in the state vector due to the measuring process does not change these properties. Since, on the other hand, every property of the system can be measured by a measuring process, it itself suggests that not only all objective properties, but, in general, all measurable properties are related to the proton. This corresponds to the usage in physics in which, entirely in the sense of classical physics, a proton is considered to be an object to which the various properties can be attributed.

Unrestricted application of a concept of substance, which derives in this manner from naive use of colloquial language within physics, leads to grave difficulties. From the exact knowledge of the physical conditions under which objective and non-objective properties can actually be measured, it follows that it is not possible to consistently relate all properties to the figurative concept of substance. This problem of so-called 'objectifiability' will be considered in more detail in the following section. The possibility of finding certain observables that remain constant with change of all properties (as for example the rest mass, charge and spin of the proton) will, on the other hand, survive. On the basis of each valid conservation law it will, in general, be possible to obtain further quantities of this kind. As we have seen, however, the existence of conservation laws is by no means sufficient to infer the existence of a substance in the classical sense. In any case, the function of the category of substance, the interpretation of different perceptions as properties of a thing, by which the cognition of objects becomes possible, is in no way given.

3. OBJECTIFIABILITY

In our considerations so far we have characterized a physical system S by its state $|\varphi\rangle$. Expanding $|\varphi\rangle$ in terms of the states $|A_i\rangle$ associated with certain properties E_{A_i} indicated only that the state $|\varphi\rangle$ occurred. We designated a certain set $E_1, E_2, \ldots E_N \ldots$ of properties as objective with respect to the state $|\varphi\rangle$. By this was meant that, with the aid of an experiment that does not change the state $|\varphi\rangle$ of the system, one can determine whether these properties occur or not. Therefore there is no difficulty in relating these objective properties to the object system in the sense of asserting that the object has these properties either in the positive or negative sense, and this is independent of whether an experimental decision has been made on this matter. For this decision does not represent a physical interference with the state of the object system, but only resolves the momentary uncertainty of the observer.

The attributing of properties to a physical object we will call 'objectification'. So far we have considered such attribution only with respect to objective properties. The question arises therefore whether all other properties, whose measurement is associated with a change of state of the system, are also objectifiable. The supposition that this is the case follows from the decomposition of the state $|\varphi\rangle$: For if one decomposes this state with respect to the two[16] orthogonal states $|E\rangle$ and $|\neg E\rangle$, which correspond to the occurrence or absence of a property E:

$$|\varphi\rangle = |E\rangle\langle E|\varphi\rangle + |\neg E\rangle\langle \neg E|\varphi\rangle,$$

then the quantities formed from the coefficients

$$w_\varphi(E) = |\langle E|\varphi\rangle|^2 \qquad w_\varphi(\neg E) = |\langle \neg E|\varphi\rangle|^2$$

are the probabilities that the property E will or will not be encountered after an E-measurement has been carried out. It can therefore be supposed that the property E can also be objectified, and that the probabilities expressed by $w_\varphi(E)$ and $w_\varphi(\neg E)$ represent a statement of the observer's knowledge, or lack thereof, for the occurrence of the property E. This supposition was already mentioned in our discussion of the measuring process, but could not be substantiated or refuted at that point. The following considerations will deal with the problem of objectifiability of arbitrary properties.

The objectification of a non-objective property poses various difficulties. The assumption that this objectification is possible leads to logical, mathematical and physical contradictions. We will investigate only the logical and mathematical paradox, but not the contradictions that arise from certain laws of physics, e.g. the energy conservation law.[17]

(1) Let us first consider the purely logical difficulties. We will assume that a property A occurs, and that the system S is in state $|A\rangle$. For a non-objective property B, which for simplicity we will assume to have only the two eigenstates $|B\rangle$ and $|\neg B\rangle$, for which the Hilbert space is thus two-dimensional, only the probabilities for the outcome of a B-measurement $w_A(B) = |\langle B|A\rangle|^2$ and $w_A(\neg B) = |\langle \neg B|A\rangle|^2$, which follow from the decomposition

$$|A\rangle = |B\rangle\langle B|A\rangle + |\neg B\rangle\langle \neg B|A\rangle$$

are known. If one further assumes that B is objectifiable, and thus if one comprehends these probabilities as a measure of the degree of knowledge that the subject has of the system in state $|A\rangle$, then it follows that these probabilities would have to be modified for any newly added information. This would be entirely in accord with the classical interpretation of probability; in statistical mechanics, for example, probability represents a statement of the subject's knowledge of an object.

Quantum theory and especially the theory of the measuring process shows, however, that the gain in new information, as for example the observation that the property B occurs, not only changes the probability for the observed property B – in this case changing it to 1 – but also for all other properties. Especially the probability of those properties that were already known is changed in the sense that the initially available knowledge of the presence of property A is transformed to a partial uncertainty that can no longer be expressed except as a probability. For after the determination that property B is present, the probability for the occurrence of property A is given by:

$$w_B(A) = |\langle A|B\rangle|^2$$

and as long as A and B are not commensurable, this value is < 1.

One would then arrive at the paradoxical result that the knowledge which the observer had of the system S – that the property A is present – is again lost and may under certain circumstances become incorrect because of the new knowledge he has gained – for example, that B occurs. This assertion is paradoxical in that it contradicts a definite proposition of classical logic. But since this proposition is independent of the remaining axioms of logic (in an axiomatic sense), the possibility still exists of accomplishing the objectification of arbitrary properties by abandoning this particular proposition without thereby relinquishing logic as a whole. The problems associated with this kind of restriction of logic will be discussed in Chapter VI. As long as one retains logic in its classical form, however, a simultaneous objectification of properties A and B is not possible.

The logical contradictions vanish immediately, however, if one relinquishes the objectification of arbitrary properties, and only relates objective properties to the system. The assertion that any particular knowledge which the subject has of the object system can be lost with the acquisition of new knowledge is thereby deprived of its paradoxical character, if it is noted that known properties cannot be changed by knowledge of other properties, but only by measurement.

It already became clear in the theory of the measuring process that knowledge acquisition relates to a situation *after* the measuring process proper. A knowledge of the properties that the system had before the measurement, is at this time no longer of any value, since these properties are no longer present. The new knowledge is available in the form of a mixture. The act of gaining knowledge in no way changes the old properties; it merely selects one of the possible new properties to be realized. It is therefore not a matter of complementarity of different bits of knowledge, but of complementarity of the properties of a physical system. The logical difficulties mentioned then do not appear.

(2) The mathematical difficulty incurred by the objectification of arbitrary properties consists in that the assumption of objectifiability leads to contradictions with the probability calculus. This problem will be illustrated with a mental experiment.

Consider a beam of particles, whose individual particles are in the state $|\varphi\rangle$. By a suitable experimental arrangement this beam is to be split into two coherent beams that correspond to the states $|B\rangle$ and $|B'\rangle$. This can be accomplished for example with a screen that has two holes (Figure 10). In every case we will assume that there are only two possible properties B and B', which are mutually exclusive, i.e. $|B\rangle$ and $|B'\rangle$ are to be orthogonal, and the Hilbert space spanned by the $|B\rangle$-vectors is to have only two dimen-

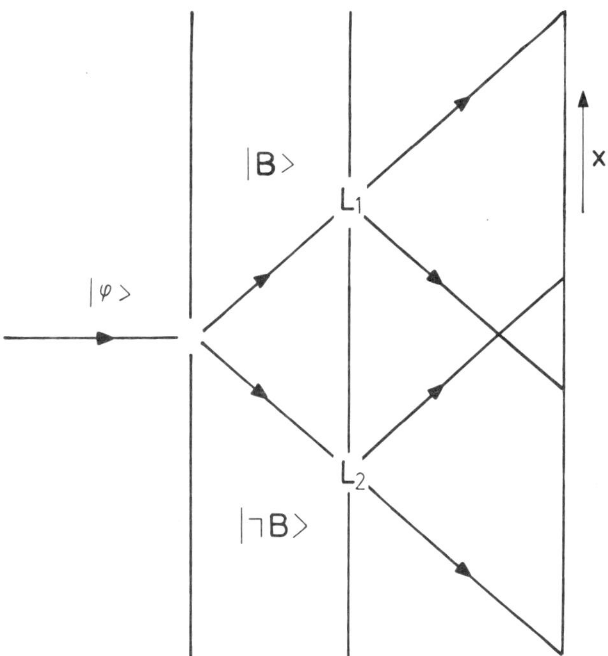

Fig. 10. Schematic representation of the two-slit experiment.

sions. Corresponding to the terminology previously introduced we designate the property B' by not-B or by $\neg B$.

The decomposition of the state $|\varphi\rangle$ into the two states $|B\rangle$ and $|\neg B\rangle$ then gives[18]

$$|\varphi\rangle = |B\rangle\langle B|\varphi\rangle + |\neg B\rangle\langle \neg B|\varphi\rangle$$

and the probabilities

$$w_\varphi(B) = |\langle B|\varphi\rangle|^2$$

$$w_\varphi(\neg B) = |\langle \neg B|\varphi\rangle|^2 .$$

The impinging particles are to be registered on a photographic plate mounted behind the screen with the two holes L_1 and L_2. We are interested only in the distribution along the vertical axis, which we denote the x-axis. If the particle strikes the photographic plate at position x, we will say that it has the property E_x and that it is in state $|x\rangle$. Since the properties E_x are not objective with respect to $|B\rangle$ and $|\neg B\rangle$, decomposition are obtained in terms of E_x, i.e. we simply decompose with respect to the observable x with the eigenvalues x[19]:

$$|\varphi\rangle = \int dx\,|x\rangle\langle x|\varphi\rangle$$

$$|B\rangle = \int dx\,|x\rangle\langle x|B\rangle$$

$$|\neg B\rangle = \int dx\,|x\rangle\langle x|\neg B\rangle.$$

Corresponding to this decomposition we introduce the probabilities for finding the position x:

$$w_B(x) = |\langle x|B\rangle|^2 \qquad w_{\neg B}(x) = |\langle x|\neg B\rangle|^2$$

$$w_\varphi(x) = |\langle x|\varphi\rangle|^2$$

We shall now assume that all properties, without exception, can be objectified. In the present case this implies that the property B and the positional properties E_x can be related to the system in the state $|\varphi\rangle$ in the sense that these properties — since nothing precise is known about their occurrence — can at least potentially be attributed to the system. The probabilities $w_\varphi(B)$, $w_\varphi(x)$, $w_B(x)$, and $w_{\neg B}(x)$ are then statements of the information that the observer has about the respective system. Whether the properties E_x and B are present or not is established objectively, but is undecided subjectively.

With these assumptions one can then argue as follows: $w_\varphi(x)$ is the probability for the occurrence of the position x, if the system has the property E_φ; $w_B(x)$ is the probability for the occurrence of the position x, if the system has the property B; $w_\varphi(B)$ is the probability that the property B will be found if the system has the property E_φ. The probability of finding both property E_x and B, (E_x, B), is therefore

$$w_\varphi(E_x, B) = w_\varphi(B) \cdot w_B(E_x)$$

and similarly

$$w_\varphi(E_x, \neg B) = w_\varphi(\neg B) \cdot w_{\neg B}(E_x).$$

The validity of these two formulae can be tested directly by experiment. If, for example, we close here L_2 so that only particles with the property B impinge on the photographic plate, then $w_\varphi(E_x, B)$ is the quantum-mechanically correct value for the probability of finding a particle at position x. A corresponding statement is valid for $w_\varphi(E_x, \neg B)$ if L_1 is closed so that all impinging particles have the property $\neg B$. Since the two cases B and $\neg B$ exhaust all possibilities, the probability of finding E_x, independently of whether B or $\neg B$ occurs, is given by:

(I) $$w_\varphi(E_x) \leqslant w_\varphi(E_x, B) + w_\varphi(E_x, \neg B).$$

This relation follows directly from probability theory.[20] If, on the other hand, one starts directly with the decomposition of $|\varphi\rangle$ with respect to the states $|x\rangle$ and $|B\rangle$, then, because of

$$w_\varphi(E_x) = |\langle x|\varphi\rangle|^2 = |\langle x|(\langle B|\varphi\rangle|B\rangle + \langle \neg B|\varphi\rangle|\neg B\rangle)|^2$$

and the previously given expressions for $w_\varphi(E_x, B)$ and $w_\varphi(E_x, \neg B)$, one obtains the result:

(II) $$w_\varphi(E_x) = w_\varphi(E_x, B) + w_\varphi(E_x, \neg B) + w_\varphi^{int}$$

where

$$w_\varphi^{int} = 2\operatorname{Re}\langle x|B\rangle\langle B|\varphi\rangle\langle\varphi|\neg B\rangle\langle\neg B|x\rangle$$

Thus for $w_\varphi^{int} > 0$, Equation (I), derived from probability-theoretical considerations, stands in direct contradiction to Equation (II), which follows from quantum-mechanical calculations. Since the additional term w_φ^{int} that appears in (II) contains the terms that are responsible for the interference of the particles of the two beams, the circumstance expressed by Equation (II) can be designated as 'interference of the probabilities'. By this is not meant, of course, that probabilities can interfere in any sense.

As we see, the assumption that arbitrary properties are objectifiable, together with the quantum-theoretical Equation (II), lead to a contradiction with the circumstance expressed in Equation (I). Since (I) can be directly deduced from probability theory, it follows that the assumed objectifiability of arbitrary properties is in contradiction with this probability-theoretical proposition. To what extent it is possible to insist on objectification by abandoning this proposition will be discussed in Chapter VI. For this purpose it will be necessary to give closer consideration to the foundations of probability theory and its connection with logic.

These considerations thus show that arbitrary properties cannot in general be objectified unless the probability theory is modified. Before the measuring process proper the quantities $w_\varphi(B)$ and $w_\varphi(\neg B)$ that appear in the decomposition of the state $|\varphi\rangle$ with respect to the property B cannot be interpreted as probabilities that express the subjective lack of knowledge, since these quantities would then have to obey different laws. The question whether a system in state $|\varphi\rangle$ has, or does not have the property B can therefore not be answered, not even by probabilities.

The contradictions with probability theory vanish immediately, however, if the quantities $w_\varphi(B)$ and $w_\varphi(\neg B)$ are consistently interpreted as probabilities that relate to a situation that must first be realized by a measuring process. Only by the measuring process and its associated change of state may the system enter one of the states $|B\rangle$ or $|\neg B\rangle$. The question, which of these two states is actually realized, is then answered by the statistical propositions of quantum theory.

As long as the system is in state $|\varphi\rangle$, the question, whether B or $\neg B$ is present, is not subjectively, but objectively undecided. Only for a certain ensemble of properties is it objectively determined whether the respective property occurs or not, so that any lack of knowledge is of purely subjective nature. Without further justification we have called these properties objective. The special circumstances just mentioned explain this terminology.

The interpretation of quantum theory presented here, which is based on the non-objectifiability of arbitrary propositions and on the theory of measurement, is apparently questioned by a mental experiment expounded by Einstein, Podolsky and Rosen (EPR) in 1935.[21, 22] Actually a detailed analysis of this experiment in the framework of quantum theory shows[23], however, that our present interpretation is completely sufficient also to explain this EPR experiment – but, that we are concerned here with an especially extreme situation[24, 25] for which the phenomenon of non-objectifiability essentially loses its meaning. Therefore it is not required by the EPR experiment that we extend the quantum-mechanical formalism due to the incorporation of hidden parameters[26, 27, 28, 29], nor is it necessary to improve the formalism in any other way.[22, 30] The interpretation of quantum theory which we have used here is sufficient to explain any experimental situation known up to now.

4. CRITIQUE OF THE CONCEPT OF SUBSTANCE IN QUANTUM THEORY

In the preceding sections we discussed in which domains of reality classical and quantum-theoretical concepts of substance can be used. If we summarize the results obtained, then the following pattern emerges:

The traditional concept of substance, especially as made precise by Kant, is characterized by the following properties:

(*a*) All properties of a thing can be related to a substance, in the sense that they can be regarded as the accidents of this substance.

(*b*) The substance persists with time, i.e. regardless of the change of the accidents, the substance is unchanged.

(*c*) The persistence of substance is not to be understood in the sense of a conservation law, for it is the immutability of a substance with time that defines what a temporal change is.

It becomes apparent from the function, which this concept of substance plays in the ordering of perceptions into objects of experience, that this concept can be applied to all things in our experience. A 'thing' is to be regarded as a completely determined content of experience, i.e. of all possible properties, the property itself or its contradiction can be ascribed to a 'thing'. Since all objects of classical physics can be comprehended as 'things' in this sense – knowledge of the canonical conjugate variables p and x already determines the values of all imaginable variables –, and there is no objection to the application of the concept of substance to such objects.

Similar relations prevail in quantum theory, if it is restricted to objective properties in state $|\varphi\rangle$. The attribution of observables to the substance then consists in relating objective properties to the state $|\varphi\rangle$, which symbolizes the substance. The quantum-mechanical concept of substance characterized by $|\varphi\rangle$ then has all the properties of the traditional concept with the single restriction that not all, but only objective properties can be considered. Consolidating these objective properties with the aid of the quantum-mechanical concept of substance does not, of course, lead to 'things' in the classical sense, but to 'quantum-mechanical objects', i.e. to systems that are in a state $|\varphi\rangle$ and that can therefore not be 'completely determined'. It is obvious, however, that in the classical limiting case, for $\hbar \to 0$, when all properties become objective, the quantum-mechanical and the classical concepts of substance coincide and similarly the quantum-mechanical merges into the classical concepts of thing. Table IV.1 summarizes these connections schematically:

TABLE IV.1

Concept	Planck's constant	Domain of application	Objects
substance according	$\hbar \to 0$	all E	classical objects
to Kant	$\hbar > 0$	all E	fictional objects
substance according	$\hbar \to 0$	all E	classical objects
to quantum theory	$\hbar > 0$	obj. E	quantum-mechanical objects

Comments to Table IV.1
The first column gives the respective concept of substance; the second column indicates the physical conditions under which it is used; column three gives the properties to which it is applied. The fourth column enumerates the different kinds of objects, which come about by the application of the concepts of substance to phenomena.

If, however, one wishes to relate all properties in some sense to a substance, in quantum theory as well, it becomes apparent that this is not possible. For on the basis of the physical conditions under which alone measurement results can be obtained, it is impossible experimentally to determine all properties for a system without changing the system. Therefore it is not possible to interpret the respective measurement results as accidents of a substance. Beyond this it is not even hypothetically possible to relate all properties to an object, even if one does not wish to determine them experimentally, since otherwise one is led to logical and probability-theoretical contradictions. With reference to the ensemble of all measurable properties the classical concept of substance, and the corresponding classical concept of thing, is therefore no longer applicable. If one wishes to comprehend them as carriers of all properties, physical systems can only be regarded as 'fictional objects'[31] because of the non-objectifiability of these properties.

NOTES

[1] Descartes, *Meditation II*, 11; English transl. by N. K. Smith, *Descartes' Philosophical Writings*, Macmillan, London, 1952, p. 208.

[2]   D. Hume, *A Treatise of Human Nature,* Book I, part IV, Section 6 and 3.

[3]   Immanuel Kant, *Kritik der reinen Vernunft,* 1781(A), 1787(B); English transl. by N. K. Smith, *Immanuel Kant's Critique of Pure Reason,* Macmillan, London, 1929, p. (A)182, p. 212.

[4]   Hume, *ibid.* Section 3.

[5]   Kant, p. (B)177, p. 180.

[6]   *Ibid.,* p. (B)183, p. 184.

[7]   *Ibid.,* p. (B)224, p. 212.

[8]   *Ibid.,* p. (B)232, p. 217.

[9]   *Ibid.,* p. (B)229, p. 216.

[10] *Ibid.,* p. (B)183, p. 184.

[11] *Ibid.,* p. (B)341, p. 291; also see p. (B)321, p. 279.

[12] With regard to Kant's illustrations of the proposition of substance ("A philosopher, on being asked . . .", *Ibid.,* p. (B)228, p. 215) it is suggestive to associate the conserved quantities with the 'quantum of substance' that Kant did not further characterize. In a careful investigation of this question C. F. v. Weizsäcker ('Kant's erste Analogie der Erfahrung und die Erhaltungssätze der Physik' in *Argumentationen, Festschrift für Josef König,* Göttingen 1965) showed clearly that from the point of view of modern physics, especially the concept of energy derived from the time-translation invariance of the laws of nature (in the sense of the special theory of relativity) corresponds to the 'quantum of substance' in Kant's conception.

[13] *Ibid.,* p. (B)599, p. 487.

[14] To every state $|\varphi\rangle$, however, there belong in general several conserved quantities, namely those quantities that are objective with respect to $|\varphi\rangle$ and that do not change with time. In analogy to the terminology used previously these quantities could be designated as the 'quantum of quantum-theoretical substance'.

[15] *Ibid.,* p. (B)599, p. 488.

[16] For simplicity we shall only consider a two-dimensional Hilbert space.

[17] See W. Heisenberg, *Die physikalischen Prinzipien der Quantentheorie,* p. 25. [English transl., *Physical Principles of the Quantum Theory,* Chicago, 1934].

[18] Since it is of no consequence to our problem, we disregard the fact that the state vector $|\varphi\rangle$ has to be normalized again after the particle beam has passed L_1 and L_2.

[19] We disregard the questions that relate to the fact that the position operator does not have a discrete spectrum and continue to designate the value x as eigenvalue.

[20] Formula (I) can be derived in the following way: Let E_A and E_B be possible properties of a given system, then the probabilities satisfy the equation

$$w(E_A) = w(E_A \wedge E_B) + w(E_A \wedge E_{\neg B})$$

where $E_A \wedge E_B$ means the conjunction. For quantum mechanical properties, which are represented by the projection operators P_A and P_B the projection operator $P_{A \wedge B}$ which represents $E_A \wedge E_B$ is given by

$$P_{A \wedge B} = \lim_{n \to \infty} (P_A P_B)^n$$

The properties E_x and B considered here should satisfy the equation

$$w_\varphi(E_x) = w_\varphi(\lim_{n \to \infty} (P_x P_B)^n) + w_\varphi(\lim_{n \to \infty} (P_x P_{\neg B})^n)$$

wherefrom on account of

$$w_\varphi(\lim_{n \to \infty} (P_x P_B)^n) \leqslant w_\varphi(P_B P_x P_B) = w_\varphi(E_x, B)$$

formula (I) follows immediately.

[21] A. Einstein, B. Podolsky and N. Rosen, *Phys. Rev.* **47** (1935), 777.

[22] D. Bohm and Y. Aharonov, *Phys. Rev.* **108** (1957), 1070.

[23] P. Mittelstaedt, *Z. Naturforsch.* **29a** (1974), 539.

[24] J. v. Neumann, *Mathematische Grundlagen der Quantentheorie*, Springer-Verlag, Berlin, 1932, p. 225. [English transl. by R. T. Beyer, Princeton University Press, 1955].

[25] W. H. Furry, *Phys. Rev.* **49** (1935), 393.

[26] J. S. Bell, *Physics* **1** (1964), 195.

[27] E. P. Wigner, *American Journal of Physics* **38** (1970), 1005.

[28] J. F. Clauser *et al.*, *Phys. Rev. Lett.* **23** (1969), 880.

[29] S. J. Freedman and J. F. Clauser, *Phys. Rev. Lett.* **28** (1972), 938.

[30] C. S. Wu and I. Shaknov, *Phys. Rev.* **77** (1950), 136.

[31] In the German edition of this book, the term used for 'fictional objects' was 'uneigentliche Gegenstände'. Since the word 'uneigentlich' was used in this context as a synonym for 'fiktiv', the term 'fictional' is presumably the most accurate translation.

CHAPTER V

THE CAUSAL LAW

In the discussion of the philosophical consequences of quantum theory, the assertion that the causal law has lost its validity in the atomic domain has played an important role. Moreover, this assertion appears in various forms. On the one hand, the causal law is declared to be inapplicable under the changed physical conditions, and then again it is considered to be invalid. Against these views, arguments about the *a priori* validity of the causal law have been pressed, an approach of importance in the philosophical tradition since Kant. The question arises therefore, how the *a priori* validity of the law of causality on the one hand, and the physical situation discovered by quantum theory on the other, can be appropriately accommodated. This chapter is devoted to this question.

1. THE CONCEPT OF CAUSALITY
IN PHILOSOPHY

The concept of causality goes back to Aristotelian philosophy, in which four causes were ascribed to things: The *causa materialis*, the material, e.g. the silver of a vessel; the *causa formalis*, the form of the vessel; the *causa finalis*, the purpose which the vessel should serve; and finally the *causa efficiens*, that which has produced the vessel, in this case the silversmith who created it.

For modern natural science only the *causa efficiens*, in general called the cause, has become important. The causal law, which deals with this particular cause, states that everything that occurs must have a cause, from which it then follows as an effect. The connection between cause and effect is thought to be regular so that at any time and at any place similar causes bring similar effects. In the language of modern physics, one would say that the law which relates cause and effect, must contain the time and location of an event as independent variables. It should be pointed out immediately, however, that a relation conforming to a law for events at different times and places can by no means always be taken as a causal law. An exact physical definition of this concept will be given later. The essential feature that distinguishes a causal law from an arbitrary law is that a causal law offers the possibility of making predictions about the future from the past and present.[1] The general validity of the causal law would then permit one to determine, from knowledge of the present state of the world, the state of the world at any future time — provided that the laws that describe the events causally are known. Under the assumption that these laws are the laws of Hamiltonian mechanics, Laplace developed in his *Théorie analytique des probabilités* the fiction of a demon that is sufficiently informed to know the present state of the world completely, and whose capacity for computation is sufficiently great to be able to calculate every future state of the world: All past and future world events would be completely known to this spirit.

With the founders of modern science, belief in the causal law in the form of causal relations existing between all events appears mostly with absolute certainty. Thus Leibniz maintains: "That everything is caused by a determined destiny is as certain as $3 \times 3 = 9$. For destiny consists of the interdependence of everything as in a chain, and will take place infallibly, as much so before it has occurred, as when it has occurred."[2] This consciousness of the unity of mathematics and nature, bordering on certainty, is repeatedly encountered in the work of Kepler, Galilei and Descartes.

The conviction of the validity of the causal law was first disturbed by the scepticism of Hume. Although Hume does not doubt the validity of the principle of sufficient reason, he denies its necessity. Only repeated experience of orderly processes evokes in us the idea that this order *must* be present. It is therefore an insight that is only simulated by custom. There is nothing other than this psychological basis for the principle of causality: "Objects have no discoverable connexion together; nor is it from any other principle but custom operating upon the imagination, that we can draw any inference from the appearance of one to the existence of another."[3]

Kant's analysis enters at this point. To begin with, Kant confirms Hume's thesis that it is not possible to prove the principle of causality by purely conceptual considerations: "How anything can be altered, and how it should be possible that upon one state in a given moment an opposite state may follow in the next moment — of this we have not, *a priori*, the least conception."[4]

Since the principle of causality cannot be deduced from the concepts of cause and effect alone, it is therefore not an analytical proposition *a priori*. In the Kantian sense, it must be a synthetic proposition. According to empiricism a synthetic judgement may be proved only by experience. But Kant shows that there are also synthetic judgements that must be thought of "together with their necessity", and which therefore are valid *a priori*. He includes the principle of causality among these judgements.

The *a priori* validity of a synthetic proposition must be proved in the transcendental sense. In this case, this means that causality must prove to be a condition for the possibility that things and processes can occur in our experience. Then, if there actually are experiences of things in space and time, the causal law must be *a priori* valid for these. "The principle of the causal relation in the sequence of appearances is therefore also valid for all objects of experience . . . as being itself the ground of the possibility of such experience."[5]

Kant did not completely show which role could be ascribed to causality in forming objects of experience. Several allusions can be found, however, in the chapter on the schematism and in the Second Analogy of Experience. The application of the category of causality to phenomena is possible by means of the transcendental schema of causality, and this proves to be the time order of successive perceptions. This idea is elaborated in the Second Analogy. Although the principle of causality itself is on the whole still formulated without reference to the transcendental schema, "Everything that happens . . . presupposes something upon which it follows according to a rule"[6], the proof of this principle explicitly uses reference to time. Kant argues that transitory objects can appear in our experience only if the manifold of perceptions has previously been brought into a temporal order. The rule by which the ordering occurs is the connection of cause and effect. Causality is therefore to be regarded as the category of one of the "fundamental concepts by which we think objects in general for appearances"[7].

Successive perceptions only establish that the state of a thing changes with time. But with that alone the temporal sequence of these states is not yet determined, at least not the objective sequence of the different states. For time itself cannot be perceived, and therefore cannot be used empirically to determine the sequence for the object. Instead this determination takes place from the aspect of a causal succession of events. These are to be understood as occurring in close succession, i.e. as continuously connected events. The application of the category of causality is therefore very closely connected with a definite continuity of the occurring events.

This order and consolidation of successive events with the aid of the category of causality has the consequence that temporal changes in the objects of experience occur in a causal manner. It thereby becomes possible to assign an individuality to these objects. By individuality we mean the possibility of recognizing a certain object at any time even though it may have changed its properties with time. Individuality presents no problem as long as there exists only one object or as long as several objects can be distinguished by time-independent properties. Difficulties only appear when we investigate several objects for which time-independent distinguishing features do not exist. In that case, identification of an object at different times is possible only if the law is known by which the properties of the system change, i.e. if the changes proceed causally. Since, according to our earlier comments, objects of experience obey the causal law, they are always identifiable while their properties may change.

Whereas the category of substance interprets all changes in phenomena as variations of the accidental properties of a substance, the category of causality explains these variations as causal variations. Things, whose properties change in a regular fashion, enter experience only by the application of both these aspects to the ordering of perceptions. Whereas the objectivity of the properties of a thing is a consequence of the application of the category of substance, individuality can be traced to the use of the category of causality. Both properties together constitute what is called an object in space and time: A readily identifiable object to which all possible properties can be related. If things defined in this way appear in our experience, they must *a priori* obey the law of conservation of substance and the law of causality. Kant does not assert, however, that all perceptions allow themselves to be ordered according to the experience of objects. But he does say that in case this does not occur, we can have no knowledge of things in space and time. "If each representation were completely foreign to every other, standing apart in isolation, no such thing as knowledge would ever arise. For knowledge is essentially a whole in which representations stand compared and connected."[8]

In the light of our experiences of contemporary physics, one cannot completely agree with these considerations. Even when perceptions are not amenable to ordering by the categories of substance and causality, and things and their changes cannot be brought into experience, it is still possible, with certain restrictions, to apprehend connections that can certainly have the character of knowledge. The following considerations will illustrate this with the example of 'fictional objects' in quantum theory.[9]

The *a priori* validity of the causal law was subsequently supported even by as empirically-minded a thinker as Helmholtz: "The law of causation bears on its face the character of a purely logical law, chiefly because the conclusions derived from it do not concern actual experience, but its interpretation, and hence it cannot be refuted by any possible experience."[10] This conception of causality as a methodological principle confirms Kant's interpretation of causality as category.

2. THE CONCEPT OF CAUSALITY
IN PHYSICS

In classical physics the causal law in the sense of the Second Analogy is strictly valid: The macroscopic objects that occur are uniquely characterized by the ensemble of measurable properties and, as things in space and time, should therefore be subject to the causal law. In mechanics it is shown that the ensemble of all properties can be completely described by the observables x and p. The causal law should make it possible to calculate from the values x_0 and p_0 of an object at time t_0, the values x_1, p_1 corresponding to any future time t_1. This requirement is satisfied by Hamilton's differential equations:

$$\dot{x} = \frac{\partial H}{\partial p} \qquad \dot{p} = -\frac{\partial H}{\partial x},$$

These equations make it possible to calculate from the state $x(t_0)$, $p(t_0)$ of a system at t_0 the state $x(t)$, $p(t)$ at any future time.

As a consequence of this conformity with natural law, the objects of classical mechanics can be regarded as identifiable objects. For if at any time t_0 several objects are given that resemble each other in all invariant characteristics – such as mass points of equal mass – and only differ with regard to the coordinates $x_1^0, p_1^0; x_2^0, p_2^0; \ldots$ then these objects can be numbered according to the coordinates x_i^0, p_i^0 for example.[11]

Because of the invariances of mechanics, all of these objects can be regarded as physically equivalent. Nevertheless these objects, marked by numbering, can again be identified at any time. Hamilton's equations of motion make it possible to calculate precisely the coordinates x_1, p_1 that appear at a later time t_1, if the forces between the individual objects are known. An object found experimentally at position x_k^1 at time t_1 can therefore be identified with the object previously marked k.

Occasionally the supposition has been expressed that by causality nothing more is meant than the existence of a strict physical law, so that the causal law need never appear in physics as a separate law. The existence of strict laws of nature suffice. That is not the case, however. As the example from mechanics showed, a causal law is distinguished in that the present state of a system is sufficient in order to calculate any future state from it. It must thus be possible to make predictions for the future. This qualification selects a type of differential equation that represents an initial value problem. Furthermore, not all solutions of such differential equations will in general be suitable, so that here the principle of causality must again effect a choice.

These connections become especially clear when going from classical to relativistic mechanics. Since the speed of light is the highest possible, causal influences are no longer possible between all world-points, but only between those that are time-like. Effects must propagate with a speed less than that of light, for which reason they can only reach points within their light cones (Figure 11).

The differential equations that describe the propagation and motion of actual physical fields must therefore contain this requirement implicitly. To be sure, the relativistic field laws do not distinguish the future from the past, so that for every two possible solutions one must be selected as conforming with causality (from the advanced and retarded solutions to the

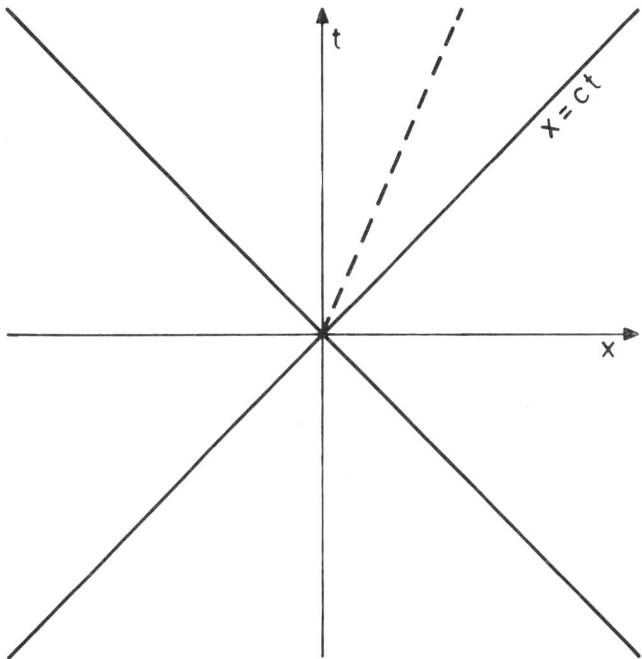

Fig. 11. Propagation of effects in the theory of relativity.

field equations). However, the principle of causality should be formulated as a separate law whenever it is not possible to describe natural processes with the aid of laws that implicitly contain the principle of causality.

Entirely analogous considerations also apply to quantum theory, provided one is restricted to objective properties. The conditions under which the experience with quantum-mechanical systems is at all possible, are the physical laws of the measuring devices, and these, in turn, are the laws of quantum theory. In this connection, we recall that it became clear in the theory of the measuring process that only certain properties of the system are measurable by cognition in the classical sense. These we there-fore designated as objective. It has previously proved possible to interpret objective properties, with the aid of the category of substance, as the accidents of a 'substance' given by the Heisenberg state. Furthermore, the changes of these accidental properties can be consolidated under causality, so that a recognizable quantum-mechanical object comes to exist through the change of phenomena.

The temporal succession of objective properties obeys a 'quantum-mechanical causal law': If at time t_0 all objective properties of a system are known, then the objective properties at subsequent times $t_1 \geqslant t_0$ can be completely calculated. The explicit form of the law which permits this calculation, is known if the Hamiltonian of the system is given. It is called the Schrödinger equation.

Corresponding to this 'quantum-mechanical causal law' one can also associate an individuality with objects characterized by a state $|\varphi\rangle$. If at time $t = t_0$, N systems are given, which have identical time-independent characteristics – several elementary particles of the same kind, for example – and which only differ with respect to their objective properties, i.e. have different states $|\varphi_1^0\rangle$, $|\varphi_2^0\rangle \ldots |\varphi_N^0\rangle$, then one can introduce an enumeration of the systems $S_1, S_2 \ldots S_N$ corresponding to these states. The systems numbered in this way can, in principle, be recognized at any time. The objective properties of all N systems can be precisely calculated for a subsequent time $t_1 \geqslant t_0$, because of the 'quantum-mechanical causal law'. Now, if a certain observable, e.g. the position x at time t_1, is an objective property of all systems and if the calculated values of these quanti-ties are $x_1^{(1)}, x_2^{(1)} \ldots x_N^{(1)}$, then a system found at the position $x_K^{(1)}$ at time t_1 can be identified with the system previously designated as S_K. The N systems must therefore be identified by observables, which at this time are objective for all systems, so that the systems can actually be distinguished by their measurement. This implies, of course, a certain restriction in the practical possibilities of identification, but in principle it causes no difficulties.

Thus the application of the causal category to time-dependent changes of the objective properties of a quantum-mechanical object has the consequence that an object results that can be recognized at any time during the changing of its properties. Furthermore, since the objective properties can be consistently interpreted as the accidents of a substance, all 'quantum-mechanical objects' that appear in experience exhibit the essential characteristics of an 'object of experience'.

The causal law, which should be *a priori* valid for all objects of experience, finds satisfactory expression for quantum-mechanical objects in the quantum-mechanical causal law. The principle of causality is in this case valid for all properties that can be consistently interpreted as accidents of a substance. A calculability of non-objective properties beyond this is not to be expected because of the *a priori* valid general causal law. As will be seen, such a calculation is in fact not possible.

3. THE INVALIDITY OF THE CAUSAL LAW
IN QUANTUM THEORY

For the objects of classical physics and for quantum-mechanical objects the causal law is strictly valid. In classical physics it appears in the form of Hamilton's equations, in quantum theory in the form of the Schrödinger equation.

As has been mentioned previously, the quantum-mechanical causal law only permits calculation of objective properties. But since, on the other hand, non-objective properties can be experimentally determined at any time, it is reasonable to look for possibilities of theoretical prediction for these properties as well. The classical causal law, which makes the calculation of all properties possible, has so far only been applied to objects of classical physics. If one now attempts to apply the causal law in its classical form to quantum-mechanical systems as well, in order possibly to be able to make predictions for non-objective properties, then two cases can arise: Either the results of quantum theory are completely taken into consideration, rendering the classical causal law not false, but inapplicable: Or one abandons quantum theory inasmuch as we assign all properties to quantum-mechanical systems, i.e. we include the non-objective properties. In this case the causal law no longer holds.

The inapplicability of the causal law is revealed when we attempt to use the classical form of this law for quantum-mechanical objects that are characterized by a state. By causal law in the sense of classical physics we understand the assertion that from a knowledge of the values x and p at time t_0, the corresponding values of x and p at any later time can be determined. The logical form of this causal law is that of an implication, whose premises P assert for an object the occurrence of certain values x_0 and p_0 at time t_0, and whose conclusion C states the calculability of the corresponding values at any later time. But in quantum theory the simultaneous knowledge of position and momentum is not possible. A situation will therefore never arise in which it can be asserted of a quantum-mechanical system that at time t_0 it is at position x_0 with momentum p_0. In the logical sense, the implication 'if P then C' is therefore to be regarded as a true position for quantum-mechanical objects too, even though the true causal law formulated in this way is quite inapplicable.

These considerations confirm once again, in a more formal way, that the general causal law is valid for all objects of experience. As noted, quantum-mechanical objects exhibit all the essential characteristics of a thing, so that the validity of the causal law is to be expected for them as well. The exact formulation of the classical causal law shows, it is true, that the previously mentioned premise P can be fulfilled for classical things in space and time, but not for quantum-mechanical objects. Consequently the classical causal law declares nothing about quantum-mechanical objects, but is nonetheless not false, of course.

In spite of these considerations it has been variously maintained that the classical causal law is no longer true in quantum mechanics. This assertion refers to a situation in which all, and thus also non-objective, properties are attributed to a quantum-mechanical system. The investigation of the category of substance has already shown that, by relating all properties to a system, logical and probability-theoretical contradictions are incurred. A similar situation prevails with respect to causality: If one regards the systems as carriers of all measurable properties, thus if one introduces fictional objects in the above sense, then the causal law becomes false.

Difficulties with regard to the *a priori* validity of the causal law asserted by Kant do not arise. As will become apparent, the 'fictional objects' discussed here lack all the essential features by which Kant's 'objects of experience' are characterized. But it is only for the latter that the causal law as a 'synthetic judgment *a priori*' has the necessary validity. There is therefore no reason to presume causal relations between fictional objects.

One is led to 'fictional objects' whenever measurement results are consistently interpreted as propositions about classical properties, with no regard for quantum theory. One thus starts from the experimental situation that every property can in principle be measured. A knowledge of quantum theory is explicitly excluded in the interpretation of experiments. The results of measurements are then attributed as objective properties to physical systems, such as electrons or atoms, without making a distinction between commensurable and incommensurable quantities. With regard to the measurement results obtained, one thus behaves precisely as in classical physics, at least as long as one does not know of quantum mechanics.

Under these circumstances the following situation can arise: For a system, the position x_0 is measured at t_0, immediately followed by a measurement of the momentum p_0. If the time-difference between the position and momentum measurements is sufficiently small[12], there is, on the assumption of constancy commonly used in classical mechanics, no hesitation in attributing to the system at time t_0 both the position x_0 and the momentum p_0. Furthermore, if the Hamiltonian of the system is known, there is no objection to calculating, with the aid of Hamilton's equations, the position x_1 and the momentum p_1 at a later time t_1. This, however, is an experimental prediction that can be tested. The result is that the values x_1 and p_1, measured at time t_1, in most instances do not agree with the calculated values. The prediction was therefore incorrect. The causal law in the form relevant to classical physics — as Hamilton's equations — is thus invalid. If one now again explicitly takes quantum theory into consideration, then it is apparent that the reason for the incorrect prediction is, of course, not the invalidity of Hamilton's equations, but the fact that position and momentum are incommensurable. The momentum measurement p_0 will completely annihilate the result of the preceding measurement of position, quite independent of the time interval between the position and momentum measurements. It is therefore not at all true that the position x_0 was present after the measurement of momentum, and consequently conclusions based on this are not sound.

The invalidity of the causal law has far-reaching consequences for the physical systems considered — and thus for 'fictional objects'. For, in contrast to the things of classical physics, these objects can no longer be individualized, not even in the restricted sense in which 'quantum-mechanical objects' can still be individualized. This is easily elucidated in the following way: If at time t_0 the positions $x_1^0, x_2^0 \ldots x_N^0$ of N otherwise identical systems are given, then nothing is known about the corresponding momenta $p_1^0, p_2^0 \ldots p_N^0$, since the p_i^0 are not objective quantities. At a later time $t_1 \geqslant t_0$ it will therefore not be possible to say where the particles $1, \ldots N$ are to be found. Now measurements of position are always associated with a perturbation of the system. If in a measurement of position one obtains $x_1^1, x_2^1 \ldots x_N^1$, it is then not possible to correlate particles determined in this way with those measured earlier. The reason for this is that time-dependent changes of fictional objects are no longer described by the causal law, but only by probabilities.

Fictional objects are thus no longer identifiable during a change of their properties. Since they can no longer be objectified, as was shown in our discussion of the concept of substance, they lack all essential features that characterize a thing in the classical sense. This justifies the comment made previously that fictional objects are not 'objects of experience', and that therefore the *a priori* validity of the causal law asserted by Kant cannot be applied to them.

That in spite of this restriction it is nevertheless meaningful under certain circumstances to deal with fictional objects as occasioned by experimental constraints. The classical objectifying interpretation of measurement results is convenient because results always appear in classical form, e.g. the deflection of a pointer of a macroscopic instrument.[13] Any incommensurability of two quantities therefore does not become immediately noticeable in the experimental results. There is thus no reason to interpret these quantities differently.

However, as has been noted, the objectifying interpretation of quantum-mechanical measurement results leads to the loss of the validity of the causal law. In interpreting measurements on a quantum-physical system in a classical way, the relinquishing of causality is not the only difficulty that ensues. This has already become apparent in connection with objectifiability, and will emerge more clearly in the discussion of quantum logic.

Summarizing our results concerning the validity of the causal law in different domains of reality, we obtain the following picture: The classical causal law is a synthetic judgement that is valid *a priori* for all objects that appear in our experience. By 'object of experience' we comprehend an object all of whose properties can be objectified without contradiction and which can be recognized as such even though its forms of appearance may change. Consequently the causal law relates to objects of classical physics and to quantum-mechanical objects, since neither objectification nor individualization is a problem for either of these two classes of objects.

A precise investigation then shows that, for these two kinds of objects, the causal law is actually valid in the form of Hamilton's equations of motion for classical objects, and in the form of the Schrödinger equation for quantum-mechanical objects. The first two items of Table V.1 relate to this result.

If one attempts to apply the classical causal law, which would make possible the calculation of all properties, to quantum-mechanical objects as well, then it turns out that the classical causal law becomes not invalid, but rather inapplicable (item 3). Fictional objects lack all essential characteristics of an 'object of experience'. They can be neither objectified nor individualized. Consequently it cannot be expected that the causal law is valid for these objects. Since fictional objects have all properties, the classical causal law is applicable to them, but it turns out to be false.

TABLE V.1

Object	Properties	Causal law	Validity
1. Classical	all properties	classical	w, a
2. Quantum-mechanical	objective	quantum-mechanical and classical	w, a $w, \neg a$
3. Fictional	all	classical	$\neg w, a$

Comments to Table V.1
The first column lists the objects, the second column lists the properties that can be attributed to the corresponding objects, the third column gives the causal law obeyed by the object. In the fourth column w means 'true', a 'applicable', $\neg w$ 'not true' and $\neg a$ 'not applicable'.

4. THE PROBLEM OF HIDDEN PARAMETERS

Both interpretations of the situation just discussed, namely that the causal law is inapplicable or false, are merely two different ways of expressing one and the same physical circumstance: From a knowledge of the state $|\varphi(t_0)\rangle$ at time t_0 it is not possible to predict the measurement values of all observables at a later time t_1. Although one can always determine the state $|\varphi(t_1)\rangle$ at time t_1, it is nevertheless not possible to calculate from it the measurement values for all observables. The reasons for this were already considered in detail in the discussion of the measuring process.

Very soon after quantum theory was established the suspicion was expressed that the failure to establish causal relations between measurement values at different times is to be traced back to the incompleteness of quantum theory. For it could well be that a state $|\varphi\rangle$ does not completely characterize a physical system, but rather that this state represents an average of more precisely defined states $|\varphi, \lambda\}$, which, besides $|\varphi\rangle$, depend on other variables λ, and by which the measurement value of all observables θ is determined for each case. These hypothetical states $|\varphi, \lambda\}$, with respect to which the measurement values of observables are no longer dispersive, are called *dispersion-free states*; the parameters λ, which determine the measurement values of the observables although they are not directly measurable themselves, are called *hidden parameters*.

The connection between the quantum-mechanical state $|\varphi\rangle$, the observables θ, their eigenvalues θ_i and their expectation values $\langle\varphi|\theta|\varphi\rangle$ on the one hand, and the dispersion-free states $|\varphi, \lambda\}$ and the associated measurement values $\theta(\varphi, \lambda)$ on the other, can be characterized by two requirements that derive directly from the conception of dispersion-free states and hidden parameters. If $\rho(\lambda)$ is the probability distribution of the parameter λ, then the mean value of the measurement values $\theta(\varphi, \lambda)$ over λ must agree with the quantum-mechanical expectation value, i.e.

$$\langle\varphi|\theta|\varphi\rangle = \int d\lambda \rho(\lambda)\, \theta\,(\varphi, \lambda)$$

must be valid. (*Expectation value postulate*). Furthermore, since the states $|\varphi, \lambda\}$ are themselves to be dispersion-free, the expectation value of an observable θ with respect to one of the states $|\varphi, \lambda_i\}$, which we designate by $\{\varphi, \lambda_i|\theta|\varphi, \lambda_i\}$ must agree with the measurement value $\theta(\varphi, \lambda_i)$ belonging to

φ and λ_i, i.e. it must agree with an eigenvalue θ_i. Therefore

$$\{\varphi, \lambda_i | \theta | \varphi, \lambda_i\} = \theta(\varphi, \lambda_i) = \theta_i$$

must be valid. (*Eigenvalue postulate*).

4.1. *The Problem of the Existence of Dispersion-Free States*

The question, whether dispersion-free states and hidden parameters exist which comply with the eigenvalue postulate and the expectation value postulate, has been investigated under various suppositions. For the discussion of these problems it is often convenient to consider, not a single system, but rather an ensemble U of N systems $S_1, S_2 \ldots S_N$, which are all in the same quantum-mechanical state $|l\rangle$ that belongs to the eigenvalue l of an observable L. With regard to the measurement value m of an observable M that is not commensurable with L one can say nothing in any particular case. For this reason one speaks of the dispersion of the observable M in the ensemble U. Now, if this can be ascribed to the fact that the state $|l\rangle$ does not completely describe the ensemble, then this implies that there must be 'hidden parameters', which can be different for the individual systems and by which the measurement value of M is determined in any particular case. Subensembles $U_1, U_2 \ldots$ of U, which belong to a value domain of hidden parameters so that the M-value derived from it is uniquely specified, are called 'dispersion-free ensembles'.

The practical preparation of dispersion-free ensembles by unmixing a given ensemble U is, because of the properties of the quantum-mechanical measuring process, certainly not possible. A simple method of such a division would be to divide the ensemble U according to the values which the observable M actually assumes in a measurement. But it is immediately apparent that this procedure does not lead to the desired goal. For once the ensemble U has been divided by an M-measurement into the subensembles $U_1, U_2 \ldots$ that belong to particular m-values, then the original l-values have been perturbed by this measuring process. None of the newly obtained subensembles can then still be associated with a fixed l-value, for in each of the newly obtained ensembles $U_1, U_2 \ldots$ it is now the quantity L that is subject to dispersion. By a similar consideration, a division according to the values of the quantity L would have the consequence that M, in turn, would be dispersed in every subensemble. The division of an ensemble into dispersion-free subensembles is therefore practically not possible.

The question, whether an ensemble U characterized by a state $|\varphi\rangle$ can at least mentally be divided into dispersion-free subensembles U_1, U_2 ..., depends essentially on the assumptions made on the dispersion-free ensembles. In our discussion of objectifiability, it already became evident for a simple example (the two-slit experiment) that the assumption that all properties of a system can be objectified, leads to statistical propositions that cannot be reconciled with the properties known to exist for quantum-mechanical states and expectation values, especially the phenomenon of the interference of probability amplitudes. If dispersion-free states do exist, then they determine, according to the eigenvalue postulate, the measuring values of all observables of the system characterized by such a state. Thus, as the example of the two-slit experiment showed, dispersion-free states cannot have all the properties that are attributed to ordinary quantum-mechanical states.

4.2. *The Propositions of von Neumann, Jauch and Piron*

These matters are expressed in a more general way by a proposition that was proved by J. von Neumann in 1932.[14] It is assumed that the expectation values $\{\varphi, \lambda|\theta|\varphi, \lambda\}$ with respect to the hypothetical dispersion-free states $|\varphi, \lambda\}$ have the same properties for arbitrary, and thus also for incommensurable, observables, as do the expectation values with respect to ordinary quantum-mechanical states. In particular, it is assumed that even for two incommensurable observables L and M the expectation values are additive[15], i.e.

$$\{\varphi, \lambda|L|\varphi, \lambda\} + \{\varphi, \lambda|M|\varphi, \lambda\} = \{\varphi, \lambda|L + M|\varphi, \lambda\}$$

For these assumptions von Neumann demonstrated that dispersion-free ensembles and states cannot exist.

To appraise the significance of this proposition, one must consider that the properties that are required here, dispersion-free states and ensembles − the additivity of the expectation values − in no way represent properties that must necessarily be demanded of an expectation value, but rather they represent a peculiarity of quantum-mechanical states which is difficult to understand. For incommensurable observables there correspond to the expectation values of L, M and $L + M$ three completely different measuring arrangements, whose results are connected by the required additivity. With regard to the hidden parameters, by which the classification of the sub-ensembles is performed, the specified assumptions imply that practically all properties of ordinary observables are required of the hidden parameters. The possibility therefore remains of introducing dispersion-free states and hidden parameters into quantum theory by qualifying or completely neglecting the given assumptions for states and observables.

A more generalized and incisive treatment of the proposition of von Neumann was provided in 1963 by Jauch and Piron.[16] The generalization consists in that not the usual formulation of quantum theory is assumed, but merely that the set-theoretical properties of the subspace $a, b \ldots$ of Hilbert space are used. If one considers, in the generalization of the concept of state, probability functions $w(a)$ of such sets, then the following proposition is valid: If the lattice L_q of subspaces $a, b \ldots$ is irreducible, and if for the function $w(a)$ it follows from[17]

$$w(a) = 1 \quad \text{and} \quad w(b) = 1 \quad \text{that} \quad w(a \wedge b) = 1$$

then there exist no dispersion-free states, i.e. probability functions, for which

$$w(a) = w(a)^2$$

is valid for all elements of the lattice.

In the Jauch-Piron proposition the conditions that deny the existence of hidden parameters and dispersion-free states are more rigorously developed than with von Neumann. In particular, it was clarified that it is only the lattice-theoretic and probabilistic assumptions that prevent the existence of dispersion-free states. But in spite of this significant formal improvement, no new physical insights have been gained. For even the Jauch-Piron proposition only shows that not all properties, which quantum-mechanical states may have, can be demanded of dispersion-free states. The probability-theoretical assumption used by Jauch and Piron is no more evident than the assumed additivity of expectation values of von Neumann. In both instances it is therefore desirable to replace these somewhat formal properties by other assumptions that are more readily understood with regard to their content. This is done by the proposition of Bell, to be discussed later.

4.3. *The Mental Experiment of Einstein, Podolsky and Rosen*

The desire to complete quantum theory by the inclusion of dispersion-free states and hidden parameters originally arose from a not very precisely defined dissatisfaction with the need to abandon objectifiability and causality in this theory. The von Neumann proof for the non-existence of dispersion-free ensembles and its refinement by Jauch and Piron are therefore seen as completely general, starting from relatively formal and in part abstruse assumptions. A rather convincing motivation for the endeavour to introduce hidden parameters into quantum theory was first expressed in the above mentioned mental experiment proposed by Einstein, Podolsky and Rosen in 1935.[18] We shall discuss this experiment in a simplified form provided by Bohm and Aharonov.[19] For this purpose we consider two physical systems S_1 and S_2 (e.g. two nucleons) both of which have a spin ½ and are in a singlet state ($^1S_0$) with total spin 0. If between these systems there is an interaction such that the angular momentum is conserved, then the total spin will always be conserved, even if the two systems separate into different directions. We now assume that the systems have been sufficiently far displaced in this way and that a measurement of the z-component of the spin of S_1 has led to the result $\sigma_z^{(1)} = +½$. Because of the assumed angular momentum conservation one then knows that a subsequent measurement of the z-component of S_2 will give the result $\sigma_z^{(2)} = -½$.

On the other hand we wish to assume that the entire system, consisting of S_1 and S_2, has the property of locality, which we define as follows:

(L)   *The result of the measurement of an observable of system S_2 is independent of which observable has been measured on system S_1, provided the systems S_1 and S_2 are sufficiently far apart.*

Applied to the mental experiment mentioned, the property of locality means that the measured value $\sigma_z^{(2)} = -½$ of the z-component of the spin of S_2 is independent of which of the three spin components σ_x, σ_y, or σ_z had previously been measured on S_1. 'Measurement' in the sense of the quantum-mechanical measuring process (see Chapter III) is to be understood as a material intervention into the system S_1, as with a Stern-Gerlach magnet, for example.

On the basis of quantum theory, the entire system formed from S_1 and

S_2, and here considered, thus has the property that the measurement of $\sigma_z^{(2)}$ with certainty gives the value $-\frac{1}{2}$, if a preceding measurement of $\sigma_z^{(1)}$ provided the value $+\frac{1}{2}$. Furthermore, because of locality, it has the property that for sufficiently large separation of S_1 and S_2 the measuring process performed on S_1 does not materially influence system S_2.

It can therefore be presumed that the value of $\sigma_z^{(2)}$ was already established prior to the two measurements and that therefore there must exist some physical variables — precisely the hidden parameters — by which the measurement value of $\sigma_z^{(2)}$ is determined. Under the assumption that the quantum theory is correct and that the systems considered do have the property of locality, the search for hidden, determining parameters in quantum theory can therefore be motivated in a plausible manner. Actually this reasoning is also fictitious. Therefore the existence of hidden parameters cannot stringently be concluded from quantum theory.[20]

4.4. *Bohm's Theory and the Problem of Locality*

The first suggestion that it is at all possible to complete the quantum theory by the inclusion of hidden parameters was given by David Bohm.[21] Bohm succeeded in explicitly constructing a theory of hidden parameters, which satisfies the eigenvalue postulate and the expectation value postulate, and which makes it possible to calculate the measurement values of the observables exactly, and thereby to establish a causal relation between the individual measurement values. Which of the assumptions mentioned in the proofs by von Neumann, Jauch and Piron were not met has not, however, been examined more closely.

Applied to the simple example of a system of two interacting spin-$\frac{1}{2}$ particles, the hidden parameters of the theory of Bohm are the paths $F_1(t)$ and $F_2(t)$ of the two particles, for which explicit equations of motion can be given. In ordinary quantum theory the concept of the path of a particle, on the other hand, has in general no observable meaning. To be sure, it does become apparent in Bohm's theory that because of their equation of motion an extremely non-local interaction exists between the orbits $F_1(t)$ and $F_2(t)$. This interaction creates a correlation between the two paths, which is quite independent of the separation of the two particles, and can thus act at arbitarily large distances.

In Bohm's theory it is therefore possible to establish with the aid of hidden parameters a causal relation between all measurement values of a system, if extremely non-local potentials are allowed. But if one abandons the property of locality, then, in the sense of the above argument, all motivation is lacking for at all introducing hidden parameters into quantum theory. For if one allows, in the framework of the Einstein-Podolsky-Rosen experiment, that correlations exist between systems S_1 and S_2, which are independent of separation, then there is no reason to be surprised that by the measurement of $\sigma_z^{(1)}$ the result of a subsequent measurement of $\sigma_z^{(2)}$ is determined. Then there is no need to search for hidden parameters that cause the determination of $\sigma_z^{(2)}$.

It could be a particular weakness of Bohm's theory that the introduction of hidden parameters is only possible if the property of locality is abandoned. It is therefore important to investigate the question whether under the assumption of locality the introduction of hidden parameters into quantum theory is at all possible. This problem is also of interest because, by comparison to the assumptions of the propositions of von Neumann, Jauch and Piron, the requirement of locality possesses considerably more physical significance. The question regarding the connection between locality and the possibility of introducing hidden parameters can be answered in a general way by a proposition that was proved by Bell in 1965[22] and for which a most general reason has been given by Wigner.[23] Bell's proposition states that under the assumption (L) of locality the introduction of hidden parameters, which determine the result of a single measurement, is fundamentally not possible. But since, on the other hand, the motivation for wanting to introduce hidden parameters derives from the idea of a local interaction, Bell's proposition means that precisely under those conditions under which their introduction would be interesting and apparently desirable, hidden parameters cannot be introduced.

Under these circumstances it is useful to pose the question, why the requirement (L) of locality, by which alone it appears desirable to introduce hidden parameters, is to be imposed upon the hidden parameters. In putting this postulate forward Einstein undoubtedly started with the obvious idea that the requirements of the special theory of relativity could only be met if changes in state are propagated with finite speeds $v \leqslant c$, by fields in the sense of a classical field theory.[24] This reason, however, is not cogent. In a Lorentz-invariant quantized field theory the assumption of the locality of field operators, corresponding to causality, i.e. the commutativity of space-like points, does not contradict the possibility that the change of state associated with a perturbation by a measurement propagates with a speed greater than the speed of light. The locality of the field operators merely prevents that, with the aid of such intervention in a system, signals are transmitted with a speed exceeding that of light.[25] With that the decisive argument for imposing the requirement (L) of locality on the hidden and unobservable parameters which describe the system considered, is eliminated. Therefore the significance of the statement, that this requirement cannot be fulfilled, is considerably reduced.

NOTES

[1]   The feasibility of actually making predictions for the future presupposes a certain simplicity of the laws of nature, since otherwise neither its cognition as law nor its manipulation for the calculation of future events would be practicable. Leibniz already drew attention to this matter in his *Metaphys. Abh.*

[2]   G. W. Leibniz, 'Von dem Verhängnisse', *Hauptschriften II*, p. 129.

[3]   D. Hume, *A Treatise on Human Nature* (ed. by T. H. Green and T. H. Grose), Longmans, Green and Co., 1909, I, p. 403.

[4]   Immanuel Kant, *Kritik der reinen Vernunft*, 1781(A), 1787(B); English edition of N. K. Smith, *Immanuel Kant's Critique of Pure Reason*, Macmillan, London, 1929, p. (B)252, p. 230.

[5]   *Ibid.*, p. (B)247, p. 227.

[6]   *Ibid.*, p. (A)189, p. 218.

[7]   *Ibid.*, p. (A)111, p. 138.

[8]   *Ibid.*, p. (A)97, p. 130.

[9]   A somewhat similar situation already emerged in our discussion of the theory of relativity (Chapter I). There it was shown as well that under certain circumstances it is impossible to order phenomena with the aid of Kant's categories. The ordering could then, of course, be carried out without difficulty by using the new concepts introduced by Einstein. For the 'fictional objects' of quantum physics an analogous solution is not possible.

[10]   Helmholtz, *Phys. Optik III*, p. 31; English transl. by James P. C. Southall, *Helmholtz's Treatise on Physiological Optics*, Dover, New York, 1962, III, p. 33.

[11]   Kant already draws attention to the possibility of individualizing undistinguishable objects by space points: "If an object is presented to us on several occasions but always with the same inner determinations (qualitas et quantitas) . . . (and) if it is appearance, we are not concerned to compare concepts; even if there is no difference whatever as regards the concepts, difference of spatial position at one and the same time is still an adequate ground for the numerical difference of the object." (Kant's *Kritik*, p. (B)319, p. 278.)

[12]   To meet the objection, that basically x_0 and p_0 were not measured at the same time t_0, but that a time interval, however short, intervened, and that it is not possible to say with certainty that the changes experienced by p during this time interval are

small, one can argue as follows: Suppose one selects for the measurement a point in time t_0, of which one knows that the momentum has the value p_0. This knowledge could be taken from earlier measurements, from which one simply calculates, with the aid of the Schrödinger equation, when p will be an objective property. Otherwise, p is a conserved quantity and does not depend on time. At t_0, when the momentum has the value p_0, the measurement of position is then made. The result x_0 is then exactly the measured position at time t_0. In the sense of classical physics one could then maintain that both x_0 and p_0 have been measured at the same instant.

[13] The reason why quantum-mechanical measuring devices must always have macroscopic dimensions was explained in Chapter III in connection with the discussion on the cut.

[14] J. v. Neumann, *Mathematische Grundlagen der Quantenmechanik*, Springer-Verlag, Berlin, 1932, especially pp. 162–171; English transl. by Robert T. Beyer, *Mathematical Foundations of Quantum Mechanics*, Princeton University Press, Princeton, 1955, pp. 305–328.

[15] *Ibid.*, p. 164; English translation, p. 309; condition *E*.

[16] J. M. Jauch and C. Piron, *Helv. Phys. Acta* **36** (1963), 827. C. Piron, *Helv. Phys. Acta* **37** (1964), 439.

[17] A further reduction of this condition is possible with the aid of a theorem by Gleason, which, however, we will not discuss further. See J. S. Bell, *Rev. Mod. Phys.* **38** (1966), 447.

[18] A. Einstein, B. Podolsky and N. Rosen, *Phys. Rev.* **47** (1935), 777. N. Bohr, *Phys. Rev.* **48** (1935), 696.

[19] D. Bohm and Y. Aharanov, *Phys. Rev.* **108** (1957), 1070.

[20] P. Mittelstaedt, *Z. Naturforschung* **29a** (1974), 539.

[21] D. Bohm, *Phys. Rev.* **85** (1952), 166, 180. 'Hidden Variables in the Quantum Theory', in *Quantum Theory III* (ed. by D. R. Bates), Academic Press, New York and London, 1962.

[22] J. S. Bell, *Physics* **1** (1965), 195.

[23] E. P. Wigner, *American Journal of Physics* **38** (1970), 1005.

[24] See: A. Einstein and M. Born, *Briefwechsel 1916–1955*, München 1969, p. 231. [English transl. Walter and Co., New York, 1971].

[25] S. Schlieder, *Commun. Math. Phys.* **7** (1968), 305, and in H.P. Dürr, *Quanten und Felder*, Vieweg, Braunschweig, 1971.

CHAPTER VI

LOGIC AND QUANTUM LOGIC

Strict analysis of quantum theory has shown that for certain propositions about quantum-mechanical systems some laws of logic lose their validity. This assertion is justified by pointing out that quantum mechanics is an empirically verified theory.

On the other hand, it must be emphasized logic is to be understood to comprise the laws immanent in reasoning, the validity of which is established by the evidence inherent in the laws, not needing to be verified for every special range of objects.

The question arises, therefore, of what is to be understood, under these circumstances, by the assertion that in the domain of quantum-mechanical propositions logic becomes, to some extent, invalid. To answer this question we must consider the basis for the evidence for the laws of logic as well as the results of quantum theory – problems that will be examined in this chapter.

1. FORMULATION OF THE PROBLEM

Logic is a theory that deals with those relationships between various propositions that are valid independent of the content of the respective propositions. Thus, the validity of a logical relationship does not depend on the detailed content of the actual propositions, but rather on the structure of the relationships investigated. Since Aristotle first developed its theory, logic has been considered to be the prototype of a science which does not depend on experience for its validity, and yet is applicable to experience. Aristotle did not question why the forms of argument established by him were, in fact, valid, but restricted himself to giving a systematic and comprehensive representation of the several forms of argument, the so-called syllogistic. Aristotle did, however, distinguish the forms of argument according to their perfection, in that he attributed to perfect conclusions a higher degree of reason than to others. Irrespective of this distinction, however, all of his forms of argument are to be regarded as inherently evident, even though the evidence may assume various degrees of clarity.[1]

Further development of logic in the Stoic-Megarian school and in scholasticism led, in particular, to the establishment of propositional logic, which could be combined with Aristotelian syllogistic into a uniform theory only within the framework of modern quantifier logic. The development of logic to the modern propositional and quantifier logic was substantially facilitated by the use of formal methods which were developed in connection with mathematical formalism. Especially the axiomatic method, familiar from investigations of geometry, contributed materially to the clarification of structural relationships in logic. Admittedly, the axiomatization of logic, as carried out principally in the investigations of Hilbert and his school, has awareness of the *a priori* validity of logical laws relegated to the background. At times the impression could thereby evolve that logic apparently was a contingent theory, the truth of which could not be verified other than by experience.

In order to meet such empiricism within logic it must be shown that the laws of logic, though valid for all statements encountered in experience, do not derive their validity from experience. Lorenzen provided, within the framework of his operative interpretation of logic, a basis for logic which satisfies both of these requirements.[2] This operative conception understands the laws of logic not as arbitrary assertions that are adapted to a specific subject matter, but rather as rules, the validity of which derives from reflection on the possibility of demonstrating the validity of propositions. In this manner every logical statement can be justified by its inherent evidence.

The precise conditions under which the validity of propositions can at all be proved thus determine the general laws that govern these propositions. Statements appearing in our experience must therefore necessarily obey the laws of logic. In Kant's terminology, logical statements consequently may be regarded as synthetic, *a priori* judgments, which are valid for all empirical statements and whose proof must be transcendental, that is, in this case must follow by our reflections on the conditions for the possibility of the proof of statements. But a proof such as this also clarifies that the *a priori* validity of the laws of logic refers only to statements for which the possibility of proof satisfies precisely the very conditions on which the derivation of logic was based.

We will denote as 'protological properties' those properties of propositions by which the possibility of proving the propositions is determined. Of the protological properties, which propositions must by necessity have if the laws of logic are to be valid for them, we will especially emphasize one, that of 'unrestricted availability'. Within the range of our formulation of the problem, this protological property delineates the domain of propositions for which the laws of logic necessarily have validity: they are applicable to all propositions which possess unrestricted availability.

Since nearly all propositions that are encountered in practice have this protological property of unrestricted availability, it is superfluous in general to state this property as a demarcation of the domain of validity of logic. Hence the laws of logic are usually designated, without further reservations, as inherently evident. Thus, all propositions of classical physics, for example, obey the laws of logic. An especially important domain of application is, in addition, presented by the propositions of mathematics, since propositions that are the consequence of schematic operations with figures according to definite rules also possess the mentioned protological property of unrestricted availability.

Despite the comprehensive applicability of logic to numerous scientific and practical problems, it has become apparent that logic loses its validity when applied to propositions that assert the actuality of arbitrary properties of quantum-mechanical systems. The reason for this is that these propositions no longer have the previously mentioned protological property of unrestricted availability, but rather only a 'restricted availability'.[3] This restriction arises from certain limitations on the possibility of proving quantum-physical propositions. The limitations, in turn, are caused by the general physical conditions to which observations of quantum-mechanical systems are subjected, and by which, therefore, propositions can at all be made about such systems.

The *a priori* validity of the laws of logic is, of course, in no way restricted by this recognition. For whenever propositions have the protological property of unrestricted availability then logic is *a priori* valid for these propositions. But if for some reason these properties are not present, if, for example, the propositions have only restricted availability, then there is no reason to expect the validity of logic for the propositions in question. In fact, under these conditions some of the laws of logic become incorrect, as we have mentioned.

To be sure, many of the laws that are already familiar from classical logic can be justified for propositions of restricted availability, but not all of them. The totality of these laws, which we will designate as 'quantum logic'[4], is to be considered as evident for propositions of restricted availability in the same way as is the more usual logic for propositions of unrestricted availability. Logic and quantum logic are therefore to be regarded as *a priori* valid theories which, however, refer to different types of propositions. The two types of proposition are distinguished not by their characteristics of content but by their protological properties, that is, by the possibilities that are available for the proof of the respective propositions. The possibilities of proof for propositions about quantum-physical systems — when compared to the possibilities of proof for propositions of classical physics — are limited by certain additional conditions which, in turn, are determined by the very conditions under which atomic systems can indeed be known, i.e. by the physical laws of the measuring instruments.

2. CLASSICAL LOGIC

Before inquiring into the basis of logic as a theory, we will first of all ask what indeed it is that logical statements express. For this purpose we consider arbitrary elementary statements, which we will denote by A, B, C. These can deal with scientific propositions such as 'The moon is round', or with mathematical statements as for example 'In a triangle, the sum of angles is $180°$', or with any statement of our experience. We will assume that it has been established how the elementary propositions A, B, C can be proved, i.e. that for each proposition a method is known which can be regarded as evidence for the respective proposition. Thereby a process is determined which decides whether or not a certain procedure is to be regarded as evidence for a certain proposition. In the case of our examples, the truth of the proposition about the moon could possibly be established by an astronomical measurement, that about the sum of angles, on the other hand, by geometrical evidence. The existence for each elementary proposition of such a method of proof determines whether the particular proposition can be said to be true. Moreover, for propositions about physical systems which are investigated here, it will always be possible to decide whether the proposition in question is true or false. Such propositions will be said to have a definite truth value.

From the elementary propositions A, B, C we can form connected propositions with the aid of the binary operation '$\rightarrow$' as for example, $A \rightarrow B$. This operation will be called here "material implication".[5] We will explain implication, colloquially described by 'if A then B' in the following manner: if someone asserts that $A \rightarrow B$, then he assumes the obligation, in case A can be proved, of justifying B. This explanation determines how an implication $A \rightarrow B$ is to be proved or is to be refuted. We can further clarify this in the form of a dialogue: one of the participants in the discussion (proponent P) asserts that $A \rightarrow B$, the other (opponent O) attempts to refute this assertion. Whenever the opponent demonstrates A, the proponent must in turn prove B in order to defend the assertion $A \rightarrow B$ claimed by him. If he succeeds, then he has prevailed in the dialogue, and has demonstrated that $A \rightarrow B$. Should he be unsuccessful, then his argument is refuted. To challenge $A \rightarrow B$ the opponent must first demonstrate A. For only by this means can he compel the proponent to prove B, or rather, only by this means does he get the opportunity of embarassing the proponent P by his inability to prove B.

Hence the dialogue for $A \to B$ reads schematically:

P	O
$A \to B$	A
Why A?	Proof of A
B	Why B?
Proof of B	

Here it is assumed that P has succeeded in demonstrating the material implication $A \to B$. In general, of course, this depends on the propositions A and B.

So far material implication has been introduced as an operation between elementary propositions A, B, C. We now wish to extend our approach by regarding the previous implications as new objects, which can be used in order to form iterated material implications of the form $(A \to B) \to (C \to D)$. An implication between other implications can be explained in the following manner: In asserting $(A \to B) \to (C \to D)$ the proponent undertakes to prove that $C \to D$, provided the opponent can demonstrate $A \to B$. The justification of the material implications $A \to B$ and $C \to D$ must now in turn be established in a dialogue. The material implication $A \to B$ between elementary propositions can, in general, be defended in a dialogue only for particular propositions A and B; similarly the particular content of the propositions A, B, C, D determines whether the material implications $A \to B$ and $C \to D$, and consequently the iterated material implications of the form $(A \to B) \to (C \to D)$ can be vindicated in dialogue. It is possible, however, that some statements of the form $A \to B$, or of the more complicated type $(A \to B) \to (C \to D)$, are always valid, independent of the particular propositions they enunciate. The significance of this is that a proponent can always successfully defend such a statement in dialogue, that is, within the framework of the discussion there is always a certain strategy of success for this statement. In the proof of generally valid statements one may therefore assume that elementary statements established by the opponent can always be proved by him. For in order to demonstrate the general validity of a statement proposed by P, it is necessary to investigate the circumstance most unfavourable for P: the case when O can in fact prove all proposed elementary propositions. On the other hand, P may only assert elementary propositions that have already been maintained by O in the dialogue. O can then not question these. For it could well be that P does not know the proof for an elementary proposition asserted by him, and we must not exclude this possibility. The assertion of the proponent must be valid for any elementary proposition.

Consequently, the dialogue about a certain logical statement is carried on such that P and O put forth their arguments alternately. These arguments are either attacks against statements which have been asserted earlier (in line x) by the antagonist, in which case we write (x) on the right hand side of the argument. Or, they defend an assertion which has been stated (in line y) and attacked later in the dialogue. In this case we write $\langle y \rangle$ on the right hand side of the defending argument. Attacks are rights which may be taken at any time in the dialogue; defences are commitments which must be carried out in the inverse order of the corresponding attacks, and at the latest when there is no longer a right for an attack.

As an example of a generally valid statement we will investigate the statement $A \rightarrow (B \rightarrow A)$. The dialogue reads:

P		O	
1. $A \rightarrow (B \rightarrow A)$		1. A	
2. Why A?		2. Proof of A	
3. $B \rightarrow A$	$\langle 1 \rangle$	3. B	(3)
4. Why B?		4. Proof of B	
5. A	$\langle 3 \rangle$	5. Why A?	
6. Refer to $O2$.			

The proof of proposition A enunciated by P in $P5$ has already been presented by the opponent of an earlier instance ($O2$). Hence the proponent can derive the proof of A explicitly within the framework of the dialogue by referring to $O2$. The proof of B must be provided by the opponent as well. Consequently the proponent can successfully defend the proposition $A \rightarrow (B \rightarrow A)$ for arbitrary A and B, since he has to prove neither statement A nor statement B. If on the other hand O cannot prove A, then there is no reason for proponent P to prove the material implication $B \rightarrow A$ stated in $P3$, since the statement $A \rightarrow (B \rightarrow A)$ has committed him to do this only if A has been proved.

A statement such as the iterated material implication $A \rightarrow (B \rightarrow A)$ just considered, which is always valid, irrespective of the elementary statements

A and *B* that it contains – one which therefore can always be successfully defended in discussion – is denoted as a logical statement or, more precisely, as an effective-logical statement. A logical statement is one for which there is always a certain strategy of success within the scope of the dialogue which does not depend on the content of the elementary propositions contained within it. The truth of a logical statement – in the sense of the dialogue – thus resides entirely in its form.

In order to present the totality of effective-logical statements in the most usual form, it is first necessary to define the concepts 'and' and 'or', or rather the connected propositions formed with the aid of these concepts. This shall again be done in the framework of the dialogue process, since logical statements containing 'and' and 'or' can then also be proved by suitable dialogue.

The conjunction $A \wedge B$ of two propositions A and B, colloquially denoted by '*A* and *B*', is explained as follows: In asserting the proposition $A \wedge B$ the proponent assumes the obligation of proving A as well as B if requested to do so by O. Hence a dialogue in which P successfully defends the assertion reads:

P	*O*
$A \wedge B$	Why A?
Proof of A	Why B?
Proof of B.	

On the basis of this explanation one can, for example, prove in a dialogue that $(A \wedge B) \to A$, a proposition which has general validity:

P	*O*
1. $(A \wedge B) \to A$	1. $A \wedge B$ (1)
2. Why A?	2. Proof of A
3. A <1>	3. Why A?
4. Refer to $O2$.	

The proponent can therefore provide the proof of A demanded of him by referring to the proof given in $O2$. O has committed himself to justifying A (and also B) by maintaining that $A \wedge B$. If O had not presented this assertion, and thus not challenged $P1$, he would have conceded the dialogue already in the first line.

The disjunction $A \vee B$ of two propositions A and B, commonly denoted by 'A or B', is explained in the following manner: If the proponent P asserts the proposition $A \vee B$, then he assumes the obligation of providing, on demand by O, the proof for at least one of the assertions A or B. A dialogue in which P successfully defends the proposition $A \vee B$ reads:

P	O
$A \vee B$	Why A?
–	Why B?
Proof of B.	

Here it is assumed that P can only prove the statement B, but not A. For the proof of $A \vee B$ this is adequate, however. As an example of a statement that contains $A \vee B$ and that has general validity we will prove the proposition $A \rightarrow (A \vee B)$:

P	O
1. $A \rightarrow (A \vee B)$	1. A (1)
2. Why A?	2. Proof of A
3. $A \vee B$ <1>	3. Why A?
4. Refer to $O2$.	

Here again the proponent can directly fall back on the proof of A provided in $O2$.

Having clarified by these definitions how the connected propositions $A \wedge B$ and $A \vee B$ are to be used in the framework of a dialogue, we can extend the class of elementary propositions discussed to this point by including the propositions $A \rightarrow B$, $A \wedge B$ and $A \vee B$ formed from A and B. We will now denote the elements of this extended class, i.e. the elementary propositions and the connected propositions derived by the use of $\rightarrow$, $\wedge$ and $\vee$, simply as propositions.

If a proposition C can be proved, i.e. defended in dialogue, we write $\vdash C$. It is useful in addition to the operations $\rightarrow$, $\wedge$ and $\vee$ on the set of propositions to define a relation $A \leqslant B$ between two propositions A and B by

$$A \leqslant B \rightleftharpoons \vdash A \rightarrow B$$

This relation is called 'implication' and must be clearly distinguished from the operation $A \to B$, denoted here as 'material implication'. According to its definition the relation $A \leqslant B$ between the propositions A and B holds if and only if the proposition $A \to B$ can be defended in a dialogue. If for two propositions the implications $A \leqslant B$ and $B \leqslant A$ are valid we call them equivalent and write $A = B$.

With the aid of the three connectives $\to$, $\wedge$ and $\vee$, the totality of statements which can always be successfully defended in a dialogue can be clearly summarized. We wish to represent the totality of these statements, denoted as affirmative logic, in the form of a logical calculus, that is, we will present a system of rules with the aid of which all implications of affirmative logic can be derived from a few implications which we will include in the rules as the point of departure of the calculus. For the formulae of the calculus we use combinations of the symbols $A, B, C \ldots$ with $\wedge, \vee, \to, \leqslant$ and bracket symbols. With these formulae $\alpha, \beta \ldots$ we establish rules for the derivation of implications $A \leqslant B$. For the designation of the rules we use the double arrow $\Rightarrow$ and the double comma ,,. The calculus of affirmative logic is then given as follows:

(L_1)	$A \leqslant A$	$A \vdash A$ (ref.)
(L_2)	$A \leqslant B \,,\, B \leqslant C \Rightarrow A \leqslant C$	$\{A \vdash B,\ B \vdash C\} \to A \vdash C.$ (trans)
(L_3)	$A \wedge B \leqslant A$	$A \wedge B \vdash A$
(L_4)	$A \wedge B \leqslant B$	$A \wedge B \vdash B$ ($\wedge$ E).
(L_5)	$C \leqslant A \,,\, C \leqslant B \Rightarrow C \leqslant A \wedge B$	$\{C \vdash A,\ C \vdash B\} \to C \vdash A \wedge B.$ ($\wedge$ I)
(L_6)	$A \leqslant A \vee B$	$A \vdash A \vee B$
(L_7)	$B \leqslant A \vee B$	$B \vdash A \vee B$ ($\vee$ I).
(L_8)	$A \leqslant C \,,\, B \leqslant C \Rightarrow A \vee B \leqslant C$	$\{A \vdash C,\ B \vdash C\} \to A \vee B \vdash C$ ($\vee$ E)
(L_9)	$(A \wedge (A \to B)) \leqslant B$	$A \wedge (A \to B) \vdash B$ ($\sim$ mP]
(L_{10})	$A \wedge C \leqslant B \Rightarrow C \leqslant (A \to B)$	$A \wedge C \vdash B. \to C \vdash A \to B$ ($\sim$ Def.)

The rule $\alpha \Rightarrow \beta$ states that if the implication α can be proved (dialogically), then the implication β can also be demonstrated (dialogically). The dialogic proof of a rule $\alpha \Rightarrow \beta$ will, therefore, be carried out in the following way: the implication α will be presupposed by the opponent as a hypothesis before the dialogue in line 0. The proponent has then to defend the implication β whereby he may refer to the hypothesis α which has been accepted by the opponent.

Rule (L_{10}) is of particular interest for subsequent investigations. We will therefore briefly consider the proof for this rule. (L_{10}) states that if one assumes the hypothesis $A \wedge C \leqslant B$ then the implication $C \leqslant (A \rightarrow B)$ can be demonstrated (in dialogue). This can be seen in the following way:

P	O
	$[0.\ A \wedge C \rightarrow B] \quad \sim \quad A \wedge C \leqslant B$
1. $C \rightarrow (A \rightarrow B)$	1. C (1)
2. $A \rightarrow B$ <1>	2. A (2)
3. $A \wedge C$ (0)	3. B <0>
4. B <2>	4.

In $P3$ we have indicated by (0) that the hypothesis $A \wedge C \rightarrow B$ in line 0 is attacked by $A \wedge C$. The sign <0> in $O3$ means that O is defending an attack which has been put forward against line 0. Similarly the sign <2> in $P4$ shows that P defends the attack against line 2.

To the extent that they have not already been given, the proofs of the other rules and statements will not here be considered. All proofs can be effected in a similar manner within the framework of the dialogue method of proof. The calculus of affirmative logic represented by L_1 to L_{10} is consistent and complete with regard to the class of implications that can be proved in dialogue. By this is meant that every implication derivable from $L_1 - L_{10}$ is dialogically true, and that furthermore every dialogically true statement can be derived from $L_1 - L_{10}$. The proof of these important properties cannot be given here.[5]

In the dialogic proofs of logical statements so far considered, propositions were demonstrated to have general validity, in that it was not necessary for the proponent to prove the elementary propositions occurring in statements. He could in each case fall back on proofs which were provided by the opponent. This method, of course, tacitly presupposes that statements, once proved in the course of a dialogue, continue to be valid, and are readily available for reference, in the further development of the dialogue. This 'protological' property of statements, so far always assumed, we will denote as 'unrestricted availability'. Unrestricted availability, which at first appears to be an almost trivial property of statements in general, is expressly emphasized here because the quantum-mechanical statements to be investigated later lack precisely this protological property. On the basis of these comments we can thus state that the calculus of affirmative logic, represented by $L_1 - L_{10}$, can be proved dialogically for all statements of unrestricted validity.

For the further development of affirmative logic and its extension to effective logic it is necessary to introduce the two propositions V (truth) and Λ (falsity). The use of both of these propositions within the framework of the dialogic method shall be established in such a way that V cannot be questioned by either participant of the dialogue and that whoever maintains Λ shall have lost the dialogue. From this definition it follows that the propositions

$$A \rightarrow V \text{ and } \Lambda \rightarrow A$$

can be proved in dialogue for all propositions A. Therefore we have the general validity of the implications $A \leqslant V$ and $\Lambda \leqslant A$. The implication $\Lambda \leqslant A$ expresses a principle familiar from the scholastic art of disputation as 'ex falso quodlibet sequitur'. The proof of both implications is simply:

P	O
$A \rightarrow V$	A
V	

Since V may not be challenged, P wins the dialogue.

P	O
$\Lambda \rightarrow A$	Λ

By being obligated to assert the Λ-statement, O loses the dialogue, P winning it.

Any generally provable formula can be substituted for the V-proposition. For if A is a formula that has a dialogic proof, then the properties of V allow one to prove that $A \rightarrow V$. But in addition $V \rightarrow A$ can also be demonstrated, since A can always be proved on demand. Conversely, A is a provable statement whenever the implication $V \rightarrow A$ has a dialogic proof. Since V is always valid, A can be proved with the aid of the implication $V \rightarrow A$.

Therefore a proposition A can be defended in a dialogue, i.e. $\vdash A$, if and only if the relation $V \leqslant A$ holds. As an interesting consequence of this equivalence we consider the relation between the implication '$\leqslant$' and the material implication '$\rightarrow$'. From L_9 and L_{10} it follows that

$$V \leqslant A \rightarrow B \Leftrightarrow A \leqslant B$$

This rule reconfirms in terms of the calculus of affirmative logic the dialogic definition of the relation $A \leqslant B$ by $\vdash A \rightarrow B$.

Affirmative logic includes all dialogically demonstrable implications between propositions which represent positive assertions. But in the framework of affirmative logic we do not yet have the possibility of negating a proposition, that is, of asserting that a particular statement is false. In order to incorporate this possibility, the concept of the negation of a proposition must be defined, that is, it must be established how the proof of the negation of a proposition is to occur within the framework of the dialogic method.

We wish to explain the negation of a proposition A (not-A), $\neg A$, by making use of the usual definition from the formalization of intuitionistic logic:

$$\neg A := A \rightarrow \Lambda$$

By basing negation on material implication, it has been established how formulae that contain negation can be proved in the framework of the dialogic method. The proponent who asserts a proposition $\neg A$ can prove this assertion in the following manner: Either O does not challenge the proposition; this is the trivial case. Or O questions $\neg A$ by asserting A. If the opponent loses the dialogue, which he has to conduct because of his assertion of A, then P has won the dialogue and, hence, has proved $\neg A$. If O wins the dialogue on A, however, then P has lost. For to prove $\neg A$ he would have to prove Λ, which, because of the previously established agreement, would cause him to lose the dialogue. A dialogue, in which P successfully defends a statement $\neg A$, has the form:

P	O
1. $\neg A$	1. A
2. Why A?	2. $-$

O cannot provide the evidence for A demanded of him in $O2$. As long as A is an elementary proposition, P will not, in general, be able to prove a negation $\neg A$ for all A, since it must be conceded that O may possibly provide a proof for A. For more complicated expressions, on the other hand, it is entirely feasible that P may succeed with the proof of $\neg A$.

The proof of the generally valid formula $\neg(A \wedge \neg A)$ will serve as an example:

P	O
1. $\neg(A \wedge \neg A)$	1. $A \wedge \neg A$ (1)
2. Why A?	2. Proof of A
3. Why $\neg A$?	3. $\neg A$
4. A (3)	4. Why A?
5. Refer to $O2$.	

The dialogue on $A \wedge \neg A$ beginning in $O1$ cannot be successfully conducted by the opponent. It is conceded that he has proved A in $O2$. For the assertion $\neg A$ in $O3$, O is obliged to provide a proof, however. This would consist in P losing the dialogue on A, beginning in $P4$. But P wins the dialogue by reference to $O2$. Hence O loses the dialogue on $A \wedge \neg A$, and the proponent has succeeded in justifying the proposition $\neg(A \wedge \neg A)$.

Because of the definition of negation

$$\neg A := A \to \wedge$$

it is not necessary to prove every implication that contains a negation by the dialogic method. In the calculus of affirmative logic the two laws

$(L_{11}) \qquad A \wedge \neg A \leqslant \wedge$
$(L_{12}) \qquad A \wedge C \leqslant \wedge \Rightarrow C \leqslant \neg A$

can be derived from L_9 and L_{10} for negation defined in this manner. L_{11} and L_{12} allow this definition of $\neg A$ to be reconfirmed in terms of the calculus $L_1 - L_{12}$, since it follows from these laws that

$$V \leqslant \neg A \Leftrightarrow A \leqslant \wedge$$

Hence, by the addition of L_{11} and L_{12} we can extend the calculus of affirmative logic to the calculus of effective logic. All propositions which can be proved in dialogue and which contain negations, can be successfully defended as propositions of this calculus.

The calculus of effective logic represented by laws $L_1 - L_{12}$ still does not comprise what is known as classical logic. Classical logic is obtained if one adds to the laws $L_1 - L_{12}$ of the effective logic the additional law:

$(L_{13}) \qquad V \leqslant A \vee \neg A$ *tertium non datur,* or law of the excluded middle

Within the framework of effective logic the important law

$$(L_{14}) \qquad A = \neg\neg A$$

follows from L_{13}; i.e. in the effective logic we have

$$V \leqslant A \vee \neg A \Rightarrow \neg\neg A \leqslant A$$

This rule can be proved either by means of $L_1 - L_{12}$ or by the process of discussion. If one presupposes the *tertium non datur* as a hypothesis, the implication $\neg\neg A \rightarrow A$ can be defended in a dialogue as follows:

P	O
	0. $V \rightarrow A \vee \neg A$
1. $\neg\neg A \rightarrow A$	1. $\neg\neg A$ (1)
2. V (0)	2. $\neg A$ <0>
3. $\neg A$ (1)	3.

Since, in the framework of effective logic, L_{14} follows from L_{13}, in this case it is actually superfluous. But for further considerations L_{14} will be also essential, and is therefore presented here.

The validity of L_{13} and L_{14} cannot be demonstrated by the described process of discussion, that is, these laws are valid only for propositions such which are known to be either true or false, but not for arbitrary propositions A, B, C. This does not create serious difficulties, however, since all propositions on natural phenomena considered here have definite truth values. For the discussion of physical propositions we can therefore always include the laws L_{13} and L_{14} in the rules of effective logic, that is, we can use the complete classical propositional logic. For the comparative investigation of classical logic, represented by $L_1 - L_{14}$, and quantum-logic, it is of advantage to note several structural properties of the formal system described by $L_1 - L_{14}$:

Because of L_1 and L_2, the propositions form a partially ordered set with respect to $\leqslant$. For every two elements, A and B, L_3, L_4 and L_5 cause the set to have a greatest lower bound $A \wedge B$, and L_6, L_7 and L_8 provide a least upper bound $A \vee B$, so that the structure investigated is a lattice. L_9 and L_{10} indicate that the lattice is relatively pseudocomplemented, that is, for every two elements A and B there always exists an element $A \rightarrow B$, the pseudocomplement of A relative to B. From this follows in particular the distributive property of the lattice and the existence of a unit element.

The assumption that a Λ-proposition exists, implies that the lattice has a null element. Hence, because of L_9 and L_{10}, there exists for each proposition A a pseudocomplement $A \to \Lambda$ or $\neg A$ relative to Λ. The existence of the pseudocomplement $\neg A$ is asserted by L_{11} and L_{12}. If in addition L_{13} is also valid, then $\neg A$ is a complement. The lattice of the propositions is then a complemented distributive lattice, or a Boolean lattice. L_{14} provides nothing new, since it is always valid for Boolean lattices.

3. THE LOGIC OF COMMENSURABLE
PROPERTIES

The elementary propositions of classical physics state that the variables x and p of classical systems have a definite value x_0 and p_0 or can be found in a certain interval $x_0 \leqslant x \leqslant x_1$, $p_0 \leqslant p \leqslant p_1$. To begin with, it is clearly evident how these propositions are to be proved precisely by measurement of the respective observables of the system investigated. Furthermore, these propositions have unrestricted availability in a dialogue. An assertion which has been justified by measurement on the object considered can at any time be quoted in the course of the dialogue. Hence for the propositions of classical physics, effective logic is valid. Beyond this, because of their particular structure, the *Tertium non datur* is also applicable, so that for propositions of a classical physical nature, the entire classical logic is valid.

A similar circumstance is revealed when we consider the logic of propositions that assert the existence of objective properties of a quantum-mechanical system. In order to define this propositional concept precisely, we will here use that formulation of quantum-mechanics which avails itself of the concept of property of a physical system. In order to treat all properties in a uniform manner, we introduce (compare Chapter III) the following definitions:

If the state $|\varphi\rangle$ of a system S lies in a closed linear manifold M_A of Hilbert space (which can of course consist of a single state), then we will say that the system has the property E_A. This is the case precisely when $P_A |\varphi\rangle = |\varphi\rangle$ is valid, where P_A is the operator that projects on M_A. Since P_A is a linear self-adjoint operator with the eigenvalues 0 and 1, it can be considered as the operator of an observable with the values 0 and 1. The occurrence of the eigenvalue 1 then indicates that the observed system has the property E_A. With the occurrence of the eigenvalue 0 $|\varphi\rangle$ lies in $M_A' = H - M_A$, the orthogonal component to M_A of the Hilbert space H, so that S has the property E_A'.

By a physical proposition A we will further mean the assertion that the state $|\varphi\rangle$ of the observed system lies in the subspace M_A, implying that it has the property E_A. Hence we will define A by:

$$A \rightleftharpoons P_A |\varphi\rangle = |\varphi\rangle$$

Propositions that assert a system to have objective properties E_A are all commensurable with respect to the state $|\varphi\rangle$. By this is meant that for two objective properties E_A and E_B

$$P_A P_B |\varphi\rangle = P_B P_A |\varphi\rangle$$

is always valid. In this case we will then say that A and B are commensurable relative to $|\varphi\rangle$. Further, if this relation is valid for every $|\varphi\rangle$, and thus

$$P_A P_B = P_B P_A$$

then we will denote A and B as commensurable.

Several relatively commensurable properties E_A, E_B, $E_C \ldots$ can be measured on a system in arbitrary sequence, without thereby affecting the result of the measurement. We will denote the totality of the laws of logic that are valid for all propositions $A, B, \ldots$ asserting the existence of objective properties $E_A, E_B, \ldots$, as the *logic of commensurable propositions*.

To begin with, it is clear how such propositions are to be proved. On the basis of its definition, A occurs precisely when in the measurement of the property E_A it is demonstrated that E_A is observed, i.e. that P_A has the eigenvalue 1. Furthermore, these propositions have unrestricted validity in a dialogue. In this case, as in classical physics, the determination of a particular property E_A is merely an act of knowing, so that proposition A can be cited at any time in the course of a dialogue. Hence the laws of effective logic, $L_1 - L_{12}$, are valid for commensurable propositions of quantum theory. It will become apparent that, if we take into consideration their explicit form, L_{13} and L_{14} are also valid for commensurable propositions.

The definition of a proposition $A \rightleftharpoons P_A |\varphi\rangle = |\varphi\rangle$ allows one to explicitly characterize, by projection operators or rather by closed linear manifolds of Hilbert space, propositions that are connected by $\wedge$, $\vee$, or $\neg$. The implication $A \leqslant B$, operatively defined in a dialogue, suggests here that whenever A has been demonstrated by a measurement, then B will always be found to be true by a suitable measurement. In view of the definition of proposition A it is necessary and sufficient that:

$$P_A |\varphi\rangle = P_A P_B |\varphi\rangle$$

If this relation is valid not only for the particular state $|\varphi\rangle$ in which the system S is found at that moment, but for all vectors of Hilbert space, then

$$P_A = P_A P_B.$$

Between the respective subspaces M_A and M_B there exists the relation

$$M_A \subseteq M_B$$

where $\subseteq$ indicates set-theoretical inclusion.

The conjunction $A \wedge B$, formed from two propositions A and B, corresponds, on the basis of

$$A \wedge B \rightleftharpoons P_{A \wedge B} \, |\varphi> \, = \, |\varphi>,$$

to the projection operator $P_{A \wedge B}$ or rather to the corresponding subspace $M_{A \wedge B}$. For the conjunction $A \wedge B$ laws L_3, L_4 and L_5 are valid. But these three laws already unequivocally determine the operator $P_{A \wedge B}$ and the subspace $M_{A \wedge B}$, for

$$P_{A \wedge B} = P_A \, P_B$$
$$M_{A \wedge B} = M_A \cap M_B$$

where $M_A \cap M_B$ is the intersection of the subspaces M_A and M_B.

On the basis of

$$A \vee B \rightleftharpoons P_{A \vee B} \, |\varphi> \, = \, |\varphi>$$

the disjunction $A \vee B$, formed from two propositions A and B corresponds to the projection operator $P_{A \vee B}$ and the associated subspace $M_{A \vee B}$. Laws L_6, L_7 and L_8 are valid for the disjunction. These three laws in turn clearly determine the operator $P_{A \vee B}$ and the subspace $M_{A \vee B}$, namely:

$$P_{A \vee B} = P_A + P_B - P_A \, P_B$$
$$M_{A \vee B} = M_A \cup M_B$$

where $M_A \cup M_B$ is the subspace spanned by M_A and M_B.

The special propositions $\wedge$ and V also easily allow themselves to be expressed by projection operators or subspaces. For on the basis of the implications $\wedge \leqslant A$ and $A \leqslant V$, valid for all A, it follows that:

$$P_A = P_A \, P_V \text{ hence } P_V = 1, M_V = H$$

and

$$P_\wedge = P_A \, P_\wedge \text{ hence } P_\wedge = 0, M_\wedge = O$$

where H is the Hilbert space and O is the null set.

On the basis of the definition

$$A \rightarrow B \rightleftharpoons P_{A \rightarrow B} |\varphi> = |\varphi>$$

the material implication $A \rightarrow B$, formed from the two propositions A and B, corresponds to the projection operator $P_{A \rightarrow B}$ or the subspace $M_{A \rightarrow B}$. Laws L_9 and L_{10} apply to the material implication. These two laws in turn clearly determine the projection operator $P_{A \rightarrow B}$ and the corresponding subspace $M_{A \rightarrow B}$, so that

$$P_{A \rightarrow B} = 1 - P_A + P_A P_B$$
$$M_{A \rightarrow B} = (H - M_A) \cup M_B$$

The relation $A \leqslant B \leftrightarrow V \leqslant A \rightarrow B$ between the implication and the material implication here corresponds to the fact that:

$$P_{A \rightarrow B} = 1$$

is valid precisely when $P_A = P_A P_B$.

Because of

$$\neg A \rightleftharpoons P_{\neg A} |\varphi> = |\varphi>$$

the negation $\neg A$ of a proposition A corresponds to the projection operator $P_{\neg A}$ or the subspace $M_{\neg A}$. On the basis of the definition $\neg A := A \rightarrow \Lambda$, $\neg A$ occurs precisely when $P_A |\varphi> = P_A P_\Lambda |\varphi>$, thus when $P_A |\varphi> = 0$. Hence

$$P_{\neg A} = 1 - P_A \text{ and } M_{\neg A} = H - M_A$$

where $H - M_A$ is the orthogonal part of the Hilbert space on M_A. It is also possible to determine $P_{\neg A}$ and $M_{\neg A}$ from laws L_{11} and L_{12}, without using the definition of $\neg A$.

On the basis of the representation of $A \rightarrow B$ by the projection operator $P_{A \rightarrow B} = 1 - P_A + P_A P_B$, and of $\neg A$ by $P_{\neg A} = 1 - P_A$, the equivalence

$$A \rightarrow B = \neg A \vee B$$

can now be derived, whereby the material implication $A \rightarrow B$ is related to disjunction and negation. This relation has validity not on the basis of effective logic, but only within the framework of classical logic. Here it follows from the special properties of the projection operators.

These considerations thus show that propositions $A \vee B, A \wedge B, A \rightarrow B$, and $\neg A$, constituted from two elementary propositions A and B, can be characterized by projection operators and subspaces that are constructed in a simple manner from the projection operators of the propositions A and B. The following table recapitulates these results:

A	P_A	M_A
$A \wedge B$	$P_A P_B$	$M_A \cap M_B$
$A \vee B$	$P_A + P_B - P_A P_B$	$M_A \cup M_B$
$A \rightarrow B$	$1 - P_A + P_A P_B$	$(H - M_A) \cup M_B$
$\neg A$	$1 - P_A$	$H - M_A$

Thus to every law of effective logic, characterized by $L_1 - L_{12}$, there corresponds a law of projection operators or closed linear manifolds. In addition, for quantum-mechanical propositions that can be represented by projection operators, laws L_{13} and L_{14} are also valid. For, because of

$$P_{A \vee \neg A} = P_A + P_{\neg A} - P_A P_{\neg A} = 1$$

$$L_{13} \qquad V \leqslant A \vee \neg A$$

is valid, and because of

$$P_A = (1 - (1 - P_A)) = P_{\neg\neg A}$$

$$L_{14} \qquad \neg\neg A \leqslant A$$

as well.

For quantum-mechanical propositions that can be represented by commuting projection operators the entire classical logic, which we have characterized by $L_1 - L_{14}$, is applicable. Hence the logic of commensurable propositions is the classical logic. In view of the representation of quantum-mechanical propositions by subspaces M_A of the Hilbert space, this conforms to the fact that closed linear manifolds of Hilbert space, corresponding to commuting projection operators, form a complementary, distributive lattice with respect to $\cap$, $\cup$ and $\neg$.

4. THE LOGIC OF INCOMMENSURABLE PROPERTIES

The logic of commensurable properties, considered in the previous section, is valid for all propositions that can be advanced concerning the occurrence of objective properties of a physical system. All statements that can be made on a quantum-mechanical object (that is, a system in a particular state $|\varphi\rangle$) are thereby comprehended. The logic of these propositions proved to be identical with the complete classical logic.

Essentially different circumstances transpire if we consider arbitrary, that is generally incommensurable, properties of a quantum-mechanical system. As we have already noted, the interest in incommensurable properties can be traced to the fact that for a given system every property can be measured, independent of whether it be objective or not. It seems sensible then that for the discussion of logical problems the investigation be not restricted to the class of commensurable properties, but that it be extended to the class encompassing all measurable properties. For often it will not be known whether two properties are commensurable or not, so that it is expedient to take arbitrary propositions into consideration from the start. Hence in this section the question will be considered whether the laws of logic, $L_1 - L_{14}$, developed for commensurable propositions, can be transferred, in total or at least in part, to arbitrary propositions, which would then lead to a logic of generally incommensurable properties.

The treatment of the logic of incommensurable properties can be carried out under two different aspects. Either one presupposes a knowledge of quantum theory in the sense that, given two propositions, it is known whether they are commensurable or not — in this case the logic remains valid in its full extent, but some of its laws lose their applicability when applied to incommensurable properties. Or one explicitly excludes a knowledge of quantum mechanics and hence relates all measurable properties in similar manner to a quantum-mechanical system, i.e. one introduces fictional objects — in this case some laws of classical logic become false. The laws of logic that remain valid under these circumstances then form in their totality a quantum logic.

It was assumed that propositions, for which it was possible to derive a logic with the aid of the dialogic method of proof, had a criterion of proof associated with them, and that propositions once proved, could be used without restriction in the dialogue. This last protological property we have previously denoted as 'unrestricted availability'.

It is also clear how in general the proof of a proposition is to be effected for incommensurable propositions concerning a quantum-mechanical system. For each proposition A and for the corresponding property E_A there exists a well-defined measuring process which produces for the observable E_A the value 1 precisely when A occurs. On the other hand, however, these propositions no longer have the protological property of unrestricted availability. In order to demonstrate this, we start with a pair of incommensurable propositions A and B and repeat once more the earlier dialogue for $A \rightarrow (B \rightarrow A)$:

P	O
1. $A \rightarrow (B \rightarrow A)$	1. A
2. Why A?	2. Proof of A by experiment on S
3. $B \rightarrow A$	3. B
4. Why B?	4. Proof of B by experiment on S
5. A	5. Why A?
6. –	

So far we have assumed that P can provide the proof of A in $P6$ by referring to $O2$. But for the following reason this is no longer possible. It is true that in $O2$, proposition A was demonstrated experimentally. But in $O4$ proposition B was proved for the same system; because of the incommensurability of A and B, the result of the prior measurement of A was thereby again destroyed, entirely or in part. Hence system S no longer has the property E_A. For this reason P can no longer in $P6$ refer to the experiment given in $O2$. It would be necessary to carry it out once more, which however, as we know, will in general not lead to the result A, as is asserted in $A \rightarrow (B \rightarrow A)$.[7]

It is apparent, therefore, that a proposition A, proved in one instance, may not simply be cited at a later stage of the discussion, when A is challenged once again. For, in general, in the course of the dialogue other propositions not commensurable with A will have been proved, and thus proposition A is no longer available. We will say, therefore, that propositions asserting the occurrence of properties for a quantum-mechanical system have only restricted availability. The availability is restricted in that a proposition A is only available for reference in a discussion if between the proof and its citing a proposition incommensurable with A has not been proved.

Restricted availability is therefore a protological property, which is attributed to all quantum-physical propositions that refer to the occurrence of properties of fictional objects, and which represents a constraint on the unrestricted availability previously discussed. Occasioned by the constraint on the availability of propositions, the restriction of the possibility of a proof by the dialogic method has then the consequence that, for example, the statement $A \to (B \to A)$, which is generally valid for propositions of unrestricted availability, can no longer be proved, unless it is explicitly assumed that A and B are commensurable.

A proposition A, of restricted availability, may therefore be cited in a dialogue only, when between the proof of A and its citation, propositions have been proved that are commensurable with A. The same is valid also for attacks. A material implication $A \to B$ asserted for instance by O, may be attacked by P only, if between the assertion of $A \to B$ and the attack propositions have been proved that are commensurable with $A \to B$. This rule, which represents a restriction of the possibility of proof in a dialogue, we will call the *commensurability rule*. The addition of this rule to the rules of dialogue so far mentioned has the consequence that not all dialogically provable implications can still be successfully defended in dialogue. Implications which, even after the inclusion of the commensurability rule into the dialogue, can still be successfully defended, and which are therefore still valid under the restrictions important for quantum-mechanical propositions, we will denote as quantum-dialogically provable. This concept of proof is more restricted than the concept of dialogic provability used so far; for every quantum-dialogically provable implication is also dialogically provable, but, as the above example has demonstrated, the converse is not true. In order to be able to decide in a concrete case whether a dialogically provable implication is provable quantum-dialogically as well, that is, in order to be able to use the commensurability rule in the framework of a dialogue, it must be clarified whether two propositions can be considered as commensurable. Commensurability is not a logical but rather a protological property. Without referring to the explicit form of the proposition used, i.e. to the theory of projection operators, one denotes two propositions A and B as commensurable if the properties E_A and E_B can be measured in arbitrary sequence on the investigated system without thereby influencing the result of the measurement. In the absence of further information about A and B one can therefore in general say nothing regarding their commensurability. From the definition of commensurability it is possible to deduce, however, that two arbitrary propositions A and B, between which one of the implications $A \leqslant B$ or $B \leqslant A$ is valid, are always commensurable. Later we will state still further relations which lead to the consequence of the commensurability of two propositions. In order to investigate the commensurability of compound propositions, such as of A and $A \to B$, it is necessary to return to the dialogic methods of proof for these propositions and to enquire whether the respective propositions can in fact be measured in arbitrary sequence on a system. Thus it can be shown, for example, that a proposition A is commensurable with $A \to B$, independently of B, but that in general this is not true for $B \to A$.

Onc can, of course, dispense with the introduction of special restrictions on the dialogic process if one can explicitly assume a knowledge of quantum mechanics for both participants in the dialogue. For then the possibility of relating two incommensurable properties to the same system is eliminated from the start: The change in state associated with the measurement would of course be known to the participants in the dialogue.

For A, B incommensurable, the proponent P could then defend $A \to (B \to A)$ as in the following dialogue:

P	O
1. $A \to (B \to A)$	1. A (1)
2. $B \to A$	2. B (2)
3. Why B?	3. $-$

The proposition B, maintained in $O2$, may, however, be challenged by P because it is known that A and B cannot both be asserted for the same quantum-mechanical object. The proposition $A \to (B \to A)$ is then justifiable because it is not applicable whenever A and B are not commensurable. This is a situation similar to that already encountered in the discussion on causality: If one presupposes a knowledge of quantum mechanics, then the classical law of causality is also true for fictional objects, because it is not applicable to them. However, as long as one considers quantum-mechanical systems and relates all measurable properties to these systems without constraint, that is, as long as one deals with fictional objects, the mentioned restriction of the dialogic method is mandatory, since otherwise statements such as $A \rightarrow (B \rightarrow A)$, though provable, could possibly be false, as we have seen.

Therefore if one restricts the concept of proof of an implication to quantum-logical provability, then one may ask which of the implications that can be dialogically proved independently of the elementary propositions they contain can still be proved after the inclusion of the commensurability rule, and thus for propositions of restricted availability. The totality of implications which are in general dialogically provable constitutes the affirmative, or rather effective logic. Similarly we will denote as affirmative or effective quantum logic the totality of implications which can be proved quantum-dialogically, independent of the propositions they contain. The implications demonstrated in this manner will be valid for arbitrary propositions of restricted validity.

For the discussion of the question, which laws of the affirmative logic are still valid in the affirmative quantum logic, we best proceed from the affirmative logic rendered as calculus in $L_1 - L_{10}$, and examine which of these laws can still be justified quantum-dialogically. It becomes evident that $L_1 - L_9$ are provable even when taking the commensurability law into account, that is, they are quantum-dialogically provable. As well as in ordinary logic we use for the formulae of the calculus combinations of the symbols $A, B, C, \ldots$ with $\wedge, \vee, \rightarrow$ and $\leqslant$ and bracket symbols. With these formulae $\alpha, \beta, \ldots$ again we establish rules for the derivation of implications $A \leqslant B$. For the designation of the rules we use the double arrow $\Rightarrow$ and the double comma ,,. A rule $\alpha \Rightarrow \beta$ means now that if the implication α can be proved quantum dialogically then the implication β can also be justified quantum dialogically. For the proof of a rule $\alpha \Rightarrow \beta$ the implication α will therefore be presupposed by the opponent as a hypothesis before the dialogue in line O. The proponent has then to defend the implication β (quantum-dialogically) whereby he may refer to the hypothesis α which has been accepted by the opponent. It is obvious that the referability of the hypothesis α is not restricted in any way. The restrictions of the referability (availability) which are formulated in the commensurability rule are not referred to a hypothesis.

As an example for a logical statement which can be proved even under the restrictions of the commensurability rule, we consider the modus ponens law (L_9)

$$(L_9) \qquad (A \wedge (A \rightarrow B)) \leqslant B$$

The dialogic proof of L_9 is as follows:

P	O
1. $(A \wedge (A \rightarrow B)) \rightarrow B$	1. $A \wedge (A \rightarrow B)$ (1)
2. Why A?	2. Proof of A
3. A (1)	3. $B <1>$
4. Why B?	4. Proof of B
5. $B <1>$	

A possible incommensurability of A and B obviously does not affect this dialogue.

Only with L_{10} does the dialogic proof fail because of the restrictions to which the proof of incommensurable propositions is subjected. This is readily apparent from the previously given proof of this law:

P	O
0.	0. $A \wedge C \rightarrow B$
1. $C \rightarrow (A \rightarrow B)$	1. C (1)
2. $A \rightarrow B <1>$	2. A (2)
3. $A \wedge C$ (0)	3. Why $A \wedge C$?

Proposition $A \wedge C$ asserted in $P3$ is not provable, however, because P cannot use the proofs of C and A which have been given in $O1$ and in $O2$. For the result of the measurement of C is in general destroyed by the subsequent measurement of A. Thus the situation is quite similar to the example $A \rightarrow (B \rightarrow A)$ previously discussed. In both cases, in fact, the same law is considered, though it is quantum logically unprovable.

The two laws:

$$(L_9) \qquad (A \wedge (A \rightarrow B)) \leqslant B$$

$$(L_{10}) \qquad A \wedge C \leqslant B \Rightarrow C \leqslant (A \rightarrow B)$$

are in fact necessary and sufficient for the existence of the three laws:

$$(Q_9) \qquad (A \wedge (A \rightarrow B)) \leqslant B$$

$$(Q_{10}) \qquad A \wedge C \leqslant B \Rightarrow (A \rightarrow C) \leqslant (A \rightarrow B)$$

$$(Q) \qquad C \leqslant (A \rightarrow C)$$

Whereas Q is no longer valid in quantum logic, Q_9 and Q_{10} are provable quantum-logically as well, i.e. dialogically provable under the restricting conditions mentioned. The dialogic proof of Q_{10} is then as follows:

P	O
0.	0. $A \wedge C \to B$
1. $(A \to C) \to (A \to B)$	1. $A \to C$ (1)
2. $A \to B <1>$	2. A (2)
3. $A \wedge C$ (0)	3. $B <0>$
4. $B <2>$	

Thus the proponent can actually prove B with the aid of the hypothetical formula $A \wedge C \to B$. A possible incommensurability of A and C has no effect in this case, since in the conclusion $(A \to C) \to (A \to B)$ the formula $A \to B$ is asserted under the assumption $A \to C$, and therefore proposition C is never explicitly proved in the course of the dialogue. The proposition $A \wedge C$ stated in $P3$ should still be proved by a subdialogue, which is however identical to the dialogue for the *modus ponens* law and will therefore not be repeated here.

 If we therefore replace L_9 and L_{10} by Q_9, Q_{10} and Q, then Q_9 and Q_{10} are, in general, still valid for incommensurable propositions. Q, in contrast, is only provable when A and C are commensurable. Of the laws of affirmative logic, $L_1 - L_8$, Q_9, Q_{10} and Q, only Q becomes invalid in the transition to quantum logic. The remaining laws, which we will designate as $Q_1 - Q_{10}$, are:

(Q_1) $A \leqslant A$

(Q_2) $A \leqslant B \,,, B \leqslant C \Rightarrow A \leqslant C$

(Q_3) $A \wedge B \leqslant A$

(Q_4) $A \wedge B \leqslant B$

(Q_5) $C \leqslant A \,,, C \leqslant B \Rightarrow C \leqslant A \wedge B$

(Q_6) $A \leqslant A \vee B$

(Q_7) $B \leqslant A \vee B$

(Q_8) $A \leqslant C \,,, B \leqslant C \Rightarrow A \vee B \leqslant C$

$$(Q_9) \qquad (A \wedge (A \to B)) \leqslant B$$

$$(Q_{10}) \qquad A \wedge C \leqslant B \Rightarrow (A \to C) \leqslant (A \to B)$$

If, for two propositions A and B, on the other hand, the implications

$$K(A, B) \rightleftharpoons A \leqslant (B \to A)$$

and

$$K(B, A) \rightleftharpoons B \leqslant (A \to B)$$

are valid for some reason, then all the implications of affirmative logic are applicable to these propositions. Within the framework of affirmative logic we may therefore consider the propositions A and B as commensurable. Furthermore it follows from the previously given protological explanation of commensurability together with the dialogic technique, that the relation $K(A, B)$ is symmetric and closed in respect to $\wedge$, $\vee$ and $\to$. In order to incorporate these properties of the commensurability relation into the quantum logical calculus, we have to add to the laws $Q_1 - Q_{10}$ the further rules:

$$(Q_{11}) \qquad A \leqslant (B \to A) \Rightarrow B \leqslant (A \to B)$$

$$(Q_{12}) \qquad A \leqslant (B \to A) \,,, A \leqslant (C \to A) \Rightarrow A \leqslant (B * C \to A)$$

Q_{11} expresses the symmetry of $K(A, B)$ and Q_{12} the closure property of the commensurability relation. Here the sign $B * C$ means any one of the connected propositions $B \wedge C$, $B \vee C$ and $B \to C$.

The laws $Q_1 - Q_{12}$ form a calculus, which we will designate as the calculus of affirmative quantum logic. This calculus is consistent with regard to the class of implications that are amenable to quantum dialogic proof. By this is meant that every implication derivable from $Q_1 - Q_{12}$ can be proved quantum dialogically. Furthermore the calculus $Q_1 - Q_{12}$ is complete with regard to the class of quantum dialogically provable implications, i.e. every quantum dialogically provable implication can be derived from $Q_1 - Q_{12}$.[8] We do not wish to consider the question of completeness in detail. It is essential for our consideration, however, that, already within the framework of affirmative logic, the commensurability rule leads to restrictions which, of course, exert an influence on the effective logic and on the complete quantum logic.

We augment the propositions so far considered by the addition of V and Λ, as we did for propositions of unrestricted validity. Here as well, the general validity of the implications $\Lambda \leqslant A$ and $A \leqslant V$ follows from establishing how these propositions are to be used in a dialogue. In particular, it follows from this that V and Λ are commensurable with all propositions. Therefore the relations

$$V \leqslant A \to V \qquad A \leqslant V \to A$$

$$\Lambda \leqslant A \to \Lambda \qquad A \leqslant \Lambda \to A$$

are valid for any proposition A.

As with commensurable properties, one can here derive a relation between the material implication $A \to B$ and the implication $A \leqslant B$. Because $V \leqslant A \to V$ it is possible, using Q_9 and Q_{10} to prove

$$A \leqslant B \Leftrightarrow V \leqslant A \to B$$

here as well, whereby the definition of $A \leqslant B$ within the body of affirmative quantum logic is reconfirmed. Furthermore it follows from Q_9, Q_{10}, Q_{11}, Q_{12} (together with $Q_1 - Q_8$) that the material implication $A \to B$ is uniquely defined by these statements.[9] Replacing the material implication $A \to B$ by $\wedge$ (and), $\vee$ (or) and $\neg$ (not) is here equally as impossible, however, as in the affirmative logic. On the other hand, we will see that in the complete quantum logic, which takes into consideration the definiteness of truth values of propositions, such a substitution is possible.

Negation, which makes possible the transition to effective quantum logic, can in turn be introduced with the aid of the dialogue mentioned above. The proponent who asserts a proposition $\neg A$ (not-A) can only be challenged by asserting A. In order to defend $\neg A$, P can only attack the proposition A. It is obvious from this definition that the negation $\neg A$ can be defended in dialogue if and only if the material implication $A \to \Lambda$ can be proved quantum dialogically. (The definition of $\neg A$ by $A \to \Lambda$ is well known from the formalization of the intuitionistic logic.) Using the dialogic definition of $\neg A$, the laws

$$(Q_{13}) \qquad A \wedge \neg A \leqslant \Lambda$$

$$(Q_{14}) \qquad A \wedge C \leqslant \Lambda \Rightarrow (A \to C) \leqslant \neg A$$

are valid. From Q_{13} and Q_{14} it then follows, that

$$V \leqslant \neg A \Leftrightarrow A \leqslant \Lambda$$

which reconfirms the statement mentioned above, that $\neg A$ can be defended in dialogue if and only if the material implication $A \to \Lambda$ can be proved quantum dialogically. Here however this dialogic equivalence has been derived from the calculus of affirmative quantum logic, after having included Q_{13} and Q_{14} in the rules of the calculus. From the dialogic definition of $\neg A$ and the previously given explanation of the concept of commensurability it further follows that A and $\neg A$ are commensurable, i.e.

$$A \leqslant (\neg A \to A) \qquad \neg A \leqslant (A \to \neg A)$$

and that the commensurability relation is closed in respect to the negation. Hence the additional law

$$(Q_{15}) \qquad A \to (B \to A) \Rightarrow A \to (\neg B \to A)$$

is valid, which has as a consequence the commensurability of A and $\neg A$. Furthermore, it follows from Q_{13}, Q_{14}, Q_{15} that the negation $\neg A$ is uniquely determined by these statements.

By means of Q_9, Q_{10}, Q_{11}, Q_{12} and by Q_{13}, Q_{14}, Q_{15} the additional rule

$$(Q_{16}) \qquad B \leqslant A \; ,, \; C \leqslant \neg A \Rightarrow A \wedge (B \vee C) \leqslant (A \wedge B) \vee (A \wedge C)$$

can easily be proved.[10] In respect to the formulation of the complete quantum logic in terms of lattice theory, we will designate Q_{16} as 'weak quasi-modularity'. Furthermore the commensurability of A and $\neg A$ together with Q_{14} give rise to the important implication $A \leqslant \neg \neg A$. The inversion, i.e. $\neg \neg A \leqslant A$ cannot be proved within the calculus represented by $Q_1 - Q_{16}$, just as in the effective logic $L_1 - L_{12}$. The claculus $Q_1 - Q_{16}$ will be called here the calculus of *effective quantum logic*. (Formula Q_{16} is actually superfluous, since it can be proved on the basis of $Q_1 - Q_{15}$. But for further considerations Q_{16} will be also essential and is therefore presented here.) Compared to effective logic, certain restrictions have appeared which manifest themselves in the laws Q_{10} and Q_{14} and in the symmetry and the closure properties of the commensurability relation (Q_{11}, Q_{12}, Q_{15}). In addition, $A \leqslant (B \to A)$ and $B \leqslant (A \to B)$ are valid in the event that two propositions A and B are known to be commensurable.

Beyond the laws of effective quantum logic characterized by $Q_1 - Q_{16}$ it is possible to prove the *tertium non datur* here as well

$$(Q_{17}) \qquad V \leqslant A \vee \neg A$$

since the elementary propositions considered here have the property of possessing a definite truth value, i.e. of being either true or false. Assuming the *tertium non datur* the inversion of $A \leqslant \neg\neg A$, i.e. $\neg\neg A \leqslant A$ and therefore

$$(Q_{18}) \qquad A = \neg\neg A$$

can also be proved. Namely in the framework of effective quantum logic (as in the effective logic) the rule

$$V \leqslant A \vee \neg A \Rightarrow \neg\neg A \leqslant A$$

can be proved either by a dialogue or by means of $Q_1 - Q_{16}$. For the proof $Q_9 - Q_{16}$ are particularly needed.[10] The calculus represented by $Q_1 - Q_{18}$ will be called the calculus of complete quantum logic. It should be remembered that in this calculus the laws Q_{16} and Q_{18} are not independent but follow from the others.

In the framework of the complete quantum logic, characterized by $Q_1 - Q_{18}$ one can — just as well as in the classical logic — express the material implication $A \rightarrow B$, which is defined by $Q_9, Q_{10}, Q_{11}, Q_{12}$ or by the relationship $A \leqslant B \leftrightarrow V \leqslant A \rightarrow B$ derived therefrom, explicitly by $\wedge$, $\vee$ and $\neg$. For the material implication in the quantum logic one obtains, in deviation from the classical logic

$$A \rightarrow B = \neg A \vee (A \wedge B)$$

The proof of this equivalence makes use of the weak quasi-modularity Q_{16} in connection with Q_{17} and Q_{18}.[11] In the formalism of complete quantum logic the two relations

$$K(A, B) \leftrightarrow A \leqslant (B \rightarrow A)$$

$$K(B, A) \leftrightarrow B \leqslant (A \rightarrow B)$$

used for the characterization of the commensurability, can be easily brought into connection with the property usually called commensurability. On account of

$$A \leqslant B \Leftrightarrow V \leqslant \neg A \vee (A \wedge B)$$

and (Q_{16}), (Q_{17}), (Q_{18}), the two implications $A \leqslant (B \rightarrow A)$ and $B \leqslant (A \rightarrow B)$ are necessary and sufficient for the two relations[11]

$$A = (A \wedge B) \vee (A \wedge \neg B)$$

$$B = (B \wedge A) \vee (B \wedge \neg A).$$

It should be mentioned that in the complete quantum logic it can be proved by the other laws that the commensurability relation $K(A, B)$ is symmetric, i.e.

$$K(A, B) \Leftrightarrow K(B, A).$$

Therefore one of the two relations is already sufficient for expressing the commensurability of two propositions A and B.[12]

In order to compare the calculus $Q_1 - Q_{18}$ of the complete quantum logic with the calculus $L_1 - L_{14}$ of the classical logic, it is useful to formulate this calculus too in terms of lattice theory. Because of Q_1, Q_2 the set of propositions A, B, . . . forms a partly ordered set with respect to the relation $\leqslant$. For every two elements A and B, Q_3, Q_4, Q_5 cause the existence of a greatest lower bound $A \wedge B$, and Q_6, Q_7, Q_8 provide the existence of a least upper bound $A \vee B$, so that the structure investigated is a lattice. If one uses the material implication, i.e. Q_9, Q_{10}, Q_{11}, Q_{12} then it follows from these rules, that for every two elements A and B there exists an element $A \rightarrow B$, which fulfills the laws Q_9, Q_{10}, Q_{11}, Q_{12}. From these laws it follows that the lattice has a unit element $A \rightarrow A$. The assumption that a proposition Λ exists means that the lattice has a null element. The laws Q_{13}, Q_{14}, Q_{15} which follow from the definition of the negation $\neg A$ imply, that for every element A there exists an uniquely defined element $\neg A$, which satisfies Q_{13}, Q_{14}, Q_{15}. On account of the *tertium non datur* Q_{17} this element $\neg A$ is a complement, and because of Q_{18} an orthocomplement. The statement Q_{16} then means that the lattice considered is an orthocomplemented, quasi-modular lattice which we will denote by L_q.

The connection of this quantum logical propositional calculus L_q with the theory of projection operators and the theory of the closed linear manifolds of the Hilbert space respectively, can again be established easily, provided one draws into consideration the explicit form of the quantum-mechanical propositions. On the basis of the definition $A \rightleftharpoons P_A |\varphi> = |\varphi>$ of a quantum-mechanical proposition, there corresponds to every proposition A a projection operator P_A or a subspace M_A. But whereas for commensurable propositions A and B the corresponding projection operators are commutable, this is, in general, no longer the case for the incommensurable propositions here considered. This has the consequence that the projection operators associated with the connected propositions $A \wedge B$, $A \vee B$, and $A \rightarrow B$ can no longer be constructed in a simple manner from the operators P_A and P_B. It is therefore no longer possible to assign to each logical statement a corresponding equation between projection operators. In contrast, the possibility of representing propositions by subspaces still remains.

The validity of an implication $A \leqslant B$ for all vectors $|\varphi>$ of Hilbert space has the consequence that $M_A \subseteq M_B$. Since a quantum-logical statement $A \leqslant B$ is always valid, independently of the respective state vector $|\varphi>$, there thus corresponds to each quantum-logically valid implication a relation $M_A \subseteq M_B$ between the corresponding subspaces. As was the case for commensurable propositions, furthermore, it is possible to clearly determine the subspaces $M_{A \wedge B}$ and $M_{A \vee B}$ of the connected propositions $A \wedge B$ and $A \vee B$ with the aid of Q_3, Q_4, Q_5 and Q_6, Q_7, Q_8, respectively, as

$$M_{A \wedge B} = M_A \cap M_B \qquad M_{A \vee B} = M_A \cup M_B$$

The subspaces of the Hilbert space therefore form a lattice L_H. For the subspace that corresponds to the negation $\neg A$ one obtains, on the basis of the definition of negation, again $M_{\neg A} = H - M_A$. For the material implication $A \rightarrow B$, which is characterized by $Q_9, Q_{10}, Q_{11}, Q_{12}$ and the relationship $A \leqslant B \leftrightarrow V \leqslant A \rightarrow B$ derived therefrom, and which can be substituted in the framework of the complete quantum logic by $\neg A \vee (A \wedge B)$ one here obtains, in a deviation from the theory of commensurable propositions:

$$M_{A \rightarrow B} = M_{\neg A} \cup (M_A \cap M_B).$$

For commensurable propositions, however, $M_{A \rightarrow B}$ does pass over to the expression $M_{\neg A} \cup M_B$ found there.

Because of this definite assignment of elementary and connected propositions to subspaces of Hilbert space, there corresponds to every relation

$M_A \subseteq M_B$, valid in the lattice of closed linear manifolds, a quantum-logical implication $A \leqslant B$. Therefore one can obtain the rules $Q_1 - Q_{18}$ of complete quantum logic also from the theory of closed linear manifolds of the Hilbert space. This is of particular interest for the rules Q_{17} and Q_{18} which cannot be derived within the calculus of effective quantum logic $Q_1 - Q_{16}$, and which are only valid, since the propositions about quantum mechanical systems can be supposed to have definite truth values.[13]

On the basis of the definition of $M_{\neg A}$ one obtains

$$M_{A \vee \neg A} = M_A \cup (H - M_A) = H = M_V$$

from which the *tertium non datur*

$$(Q_{17}) \qquad V \leqslant A \vee \neg A \qquad \text{follows.}$$

In the framework of the effective quantum logic, statement Q_{18} follows from Q_{17}. Independent of this, however, one obtains on account of

$$M_{\neg \neg A} = H - (H - M_A) = M_A$$

the statement

$$(Q_{18}) \qquad A = \neg \neg A$$

directly from the properties of subspaces of the Hilbert space. $M_{\neg A}$ is the uniquely determined subspace, which is completely orthogonal on M_A. In the lattice L_H of subspaces of Hilbert space, $M_{\neg A}$ is the orthocomplement of M_A. Therefore L_H is an orthocomplemented lattice.

For the closed linear manifolds of Hilbert space, besides orthocomplementarity, the relation of quasimodularity

$$M_A \subseteq M_B, M_C \subseteq M_{\neg B} \Rightarrow M_B \cap (M_A \cup M_C) = M_A$$

is valid and corresponds to the quantum-logical statement[10]

$$(\tilde{Q}_{16}) \qquad A \leqslant B,, \ C \leqslant \neg B \Rightarrow B \wedge (A \vee C) = A.$$

This statement agrees exactly with (Q_{16}) if $\neg A$ in (Q_{16}) is the orthocomplement to A. Therefore the closed linear manifolds of the Hilbert space form, in respect to the relation $\subseteq$ and the operations $\cap$, $\cup$ and $\neg$, an orthocomplemented quasimodular lattice.

In the lattice L_H of subspaces of the Hilbert space the two (symmetric) relations

$$A = (A \wedge B) \vee (A \wedge \neg B)$$

$$B = (A \wedge B) \vee (\neg A \wedge B)$$

previously used for the designation of the commensurability of two propositions, correspond to the equations

$$M_A = (M_A \cap M_B) \cup (M_A \cap M_{\neg B})$$

$$M_B = (M_A \cap M_B) \cup (M_{\neg A} \cap M_B).$$

On the other hand these equations are equivalent[16] to the condition

$$P_A P_B = P_B P_A$$

which expresses the simultaneous measurability of the observables P_A and P_B and which is often considered as a convenient expression for commensurability.

5. PROBABILITY AND QUANTUM LOGIC

Among the most important consequences of quantum logic are certain restrictions that bear on the application of probability calculus to quantum-theoretical problems. In this section we wish to compare classical and quantum-mechanical probability theory. A first indication that — on the assumption of objectifiability — several laws of probability lose their validity has already appeared in the treatment of probability interference phenomena (Chapter IV). Hence for illustrating the relation between quantum logic and the theory of probability we shall, in this section, once more consider the two-slit-experiment already discussed earlier.

5.1. *Classical Probability Theory*

To begin with, we introduce the concept of probability by taking as our point of departure a physical system S in a state $|\varphi\rangle$ and propositions A, B, C, which state the existence of properties of this system. These properties do not necessarily have to be objective; rather we will, in general, consider arbitrary measurable properties. By the concept of the probability $w(A)$, that a property E_A occurs in the system S, one intuitively understands the relative frequency with which the property E_A occurs in measurements of the property E_A on a large number of systems in the same state $|\varphi\rangle$. Similarly, by the probability for the connected propositions $A \wedge B$, $A \vee B$, $\neg A$ is to be understood the relative frequency of the occurrence of the connected propositions. From this interpretation of probability as relative frequency it follows that the probability $w(A)$ of a proposition A is a real function $0 \leqslant w(A) \leqslant 1$, where the values 0 and 1 are taken when instead of A the propositions Λ and V appear, i.e. $w(\Lambda) = 0$ and $w(V) = 1$.

In the mathematical theory of probability the concept of probability is defined independently of the frequency-based interpretation mentioned, though the definition, by way of the following axioms, is closely related:

W_1 The probability is a real, non-negative function $w(x) \geqslant 0$ which is defined for all elements x of a Boolean lattice L_B.

W_2 If $\cap$ and $\cup$ are the connecting operations of the lattice considered, and if n is the null element, then for two elements x and y, for which $x \cap y = n$, the relation $w(x \cup y) = w(x) + w(y)$ applies.

W_3 — If $x_1, x_2, x_3 \ldots$ are elements of the lattice, which satisfy the conditions $x_i \cap x_k = n$ in pairs, then $w(x_1 \cup x_2 \cup x_3 \ldots) = w(x_1) + w(x_2) + w(x_3), + \ldots$ applies.

W_4 — If $w(x) = 0$ then $x = n$.

W_5 — If e is the unit element of the lattice then $w(e) = 1$.

If we restrict ourselves to commensurable propositions A, B, C, then these form, with respect to the logical connectives $\wedge$ and $\vee$, a Boolean lattice whose null and unit elements are $\wedge$ and $\vee$ respectively. A function $w(A)$ that is defined for all propositions A of this lattice and that satisfies axioms $W_1 - W_5$ of the probability definition is therefore to be regarded as a probability in the mathematical sense. If we consider, as we have done so far, a system in a state $|\varphi\rangle$, then a function that satisfies these conditions can immediately be given. Thus

$$w_\varphi(A) = \langle \varphi | P_A | \varphi \rangle$$

is such a function, which one can therefore denote as the probability for the occurrence of the property E_A in the system investigated. We will not here enter into the proof of this fact.[17]

To the quantum-mechanical propositions A, B, C there correspond closed linear manifolds M_A, M_B, M_C of Hilbert space, or rather the associated projection operators P_A, P_B, P_C. In particular, for the domain of commensurable propositions the projection operators of the connected propositions $A \wedge B$, $A \vee B$, $\neg A$ can be formed by simple combinations of the projection operators P_A and P_B. Furthermore, an implication $A \leqslant B$ has the consequence

$$P_A | \varphi \rangle = P_B P_A | \varphi \rangle$$

from which it follows, because of the properties of projection operators, that

$$w_\varphi(A) \leqslant w_\varphi(B).$$

Every logical implication $A \leqslant B$ results therefore in an inequality between the corresponding probabilities. A logical equivalence $A = B$, in particular, corresponds to an equation

$$w_\varphi(A) = w_\varphi(B).$$

It follows, for example, from the logical equivalence

$$A = (A \wedge B) \vee (A \wedge \neg B)$$

that for the corresponding projection operators

$$P_A = P_{A \wedge B} + P_{A \wedge \neg B}$$

and for the probabilities

$$(\mathrm{I}') \qquad w_\varphi(A) = w_\varphi(A \wedge B) + w_\varphi(A \wedge \neg B)$$

where the expressions $w_\varphi(A \wedge B)$ and $w_\varphi(A \wedge \neg B)$ can in turn be resolved into the products $w_\varphi(A)w_A(B)$ and $w_\varphi(A)w_A(\neg B)$. Corresponding to $P_{\neg B} = 1 - P_B$, $w_A(\neg B)$ is given by $1 - w_A(B)$.

5.2. *Quantum-Logical Theory of Probability*

The totality of all, including incommensurable, quantum-mechanical propositions A of the form $|\varphi> \subseteq M_A$ constitutes, with regard to the operations $\wedge$, $\vee$ and $\neg$, a σ-complete, atomic, orthocomplemented and quasi-modular lattice L_q. In order to be able to introduce a probability for the propositions summarized in this lattice, we define as 'quantum-logical probability' a real function $p(A)$, which is defined for all $A \subseteq L_q$ and which satisfies the following axioms:[18]

$W_1^{(q)}$	For all elements $A \subseteq L_q$, $0 \leqslant p(A) \leqslant 1$
$W_2^{(q)}$	$p(\wedge) = 0$, $p(V) = 1$
$W_3^{(q)}$	If $\{A_i\}$ is a series with $A_i \leqslant \neg A_k$ for $i \neq k$ then

$$p(A_1) + p(A_2) + \ldots = p(A_1 \vee A_2 \vee \ldots)$$

$W_4^{(q)}$ For every finite series $\{A_i\}$ it follows from $p(A_i) = 1$ for all $A_i \subseteq \{A_k\}$ that

$$p(A_1 \wedge A_2 \wedge \ldots) = 1$$

$W_5^{(q)}$ If $A \neq \wedge$, then there exists a p with $p(A) \neq 0$

                  If $A \neq B$, then there exists a p with $p(A) \neq p(B)$

Axioms $W_1^{(q)} - W_5^{(q)}$ give a real and realistic generalization of the concept of probability determined by $W_1 - W_5$. The generalization is reasonable since for Boolean sublattices $L_B \subseteq L_q$ the probabilities $p(A)$ are in agreement with the classical probability concept defined by $W_1 - W_5$. For Boolean lattices $W_4^{(q)}$ is, of course, a consequence of the remaining axioms $W_1^{(q)}$, $W_2^{(q)}$, $W_3^{(q)}$ and $W_5^{(q)}$. On the other hand, $W_1^{(q)} - W_5^{(q)}$ is a real generalization of $W_1 - W_5$, since not all statements that are provable for $w(A)$ can be proved for $p(A)$ as well. An example of such a statement will be discussed in connection with the two-slit-experiment.

A function $p(A)$, which satisfies axioms $W_1^{(q)} - W_5^{(q)}$ and which, for commensurable properties, is in agreement with the function $w_\varphi(A)$, can be readily given. Once again

$$w_\varphi(A) = \langle\varphi|P_A|\varphi\rangle$$

is such a function. For incommensurable P_A and P_B it also holds that from

$$\langle\varphi|P_A|\varphi\rangle = \langle\varphi|P_B|\varphi\rangle = 1$$

the relation

$$\langle\varphi|P_{A \wedge B}|\varphi\rangle = 1$$

follows. Thus axiom $W_4^{(q)}$, which was newly added with respect to $W_1 - W_5$, is satisfied by $w_\varphi(A)$ as well. The functions $w_\varphi(A)$ therefore represent a reasonable generalization of probability in the case that arbitrary, not necessarily commensurable, observables are considered.

We do not here wish to deal systematically with the restrictions which emerge for the calculus of probability when one considers all measurable properties. Rather we shall investigate only a characteristic example in which the restrictions become especially conspicuous. We again choose the example previously mentioned. Whereas for commensurable A and B the equivalence

$$A = (A \wedge B) \vee (A \wedge \neg B)$$

is valid, for incommensurable propositions A and B, only the implication

$$(A \wedge B) \vee (A \wedge \neg B) \leqslant A$$

applies; the converse

$$A \leqslant (A \wedge B) \vee (A \wedge \neg B)$$

is, on the other hand, not quantum-logically provable. As has been mentioned, the relation (I') follows from the logical equivalence. From the quantum-logically valid implication

$$(A \wedge B) \vee (A \wedge \neg B) \leqslant A$$

the inequality[19]

$$w_\varphi((A \wedge B) \vee (A \wedge \neg B)) \leqslant w_\varphi(A)$$

follows. Because of $W_3^{(q)}$ and because $M_{A \wedge B} \subseteq H - M_{A \wedge \neg B}$, the left side of the inequality can be additively resolved, so that here only

$$(\tilde{\mathrm{I}}) \qquad w_\varphi(A \wedge B) + w_\varphi(A \wedge \neg B) \leqslant w_\varphi(A)$$

follows, in contrast to Equation (I').

5.3. *Interpretation*

In order to perceive the significance of the generalized probability functions $p(A)$ it should be recalled that the statements of classical probability obtained from $W_1 - W_5$ are not, under certain circumstances, compatible with the quantum-mechanical equations for the calculation of the functions $\langle \varphi | P_A | \varphi \rangle = w_\varphi(A)$. For if one objectifies all properties of a system in the sense that one assumes that these properties are or are not attributable to the respective system, then the propositions on the functions $w_\varphi(A)$ derived from quantum theory contradict the statements of classical probability theory. One can resolve this conflict if one does not undertake a simultaneous objectification of incommensurable properties. The expressions $w_\varphi(A)$ appearing in quantum-mechanical equations can then be interpreted as probabilities in the sense of $W_1 - W_5$, although only as the probabilities for the occurrence of the property A subsequent to a measurement on the system that was initially in the state $|\varphi\rangle$.

It is, however, not imperative to abandon the objectifiability of quantum-mechanical properties, since the contradiction mentioned can be resolved in a different manner. For if one objectifies all, and thus also incommensurable, properties of a quantum-mechanical system, then the propositions on the occurrence of these properties do not form a Boolean lattice of propositions, but only the lattice of propositions L_q of quantum logic to quantum logic. The functions $p(A)$ are then probabilities for the properties E_A in the sense of the relative frequency of occurrence of the property. However, the statements of the 'quantum-logical probability theory' no longer contradict the quantum-mechanical equations for the functions $w_\varphi(A)$, for which reason the objectification of all properties can be effected without contradiction if one replaces logic by quantum logic, and probability theory by quantum-logical probability theory.

Thus, even under assumption of the objectifiability of all properties, one can interpret the expressions $w_\varphi(A)$ as probabilities — in the sense of relative frequencies — if for quantum-mechanical propositions one restricts logic to quantum logic. This restriction of logic, and hence of the theory of probability, is consistent, however, insofar as the assumption of objectifiability of all properties only permits quantum logic. The contradiction between objectifiability, quantum mechanics and probability theory can also be resolved if one notes all the consequences of the assumption of objectifiability and correspondingly restricts the logic and the probability theory.

5.4. *Example*

A simple experimental arrangement in which the restrictions on the probability calculus find immediate expression is the two-slit-experiment already mentioned in connection with objectifiability. A beam of particles is divided into two by a screen containing two slits (Figure 12). Behind the screen the particles are registered on a photographic plate. Let the propositions B and $\neg B$ correspond to a particle passed through the upper and the lower slit, respectively; by A we will mean the proposition that the particle has impinged on a certain position x on the photographic plate, that is, after its registration on the plate the particle is in state $|x\rangle$. The states that are associated with the two propositions B and $\neg B$ we will call $|B\rangle$ and $|\neg B\rangle$, respectively; and the state in which the particle found itself at the beginning of the experiment, we denote as $|\varphi\rangle$. From the decomposition

$$|\varphi\rangle = |B\rangle\langle B|\varphi\rangle + |\neg B\rangle\langle \neg B|\varphi\rangle$$

$$|B\rangle = \int dx |x\rangle\langle x|B\rangle$$

$$|\neg B\rangle = \int dx |x\rangle\langle x|\neg B\rangle$$

the probabilities can be obtained

$$w_\varphi(B) = |\langle B|\varphi\rangle|^2 \qquad w_\varphi(\neg B) = |\langle \neg B|\varphi\rangle|^2$$

$$w_B(A) = |\langle x|B\rangle|^2 \qquad w_{\neg B}(A) = |\langle x|\neg B\rangle|^2$$

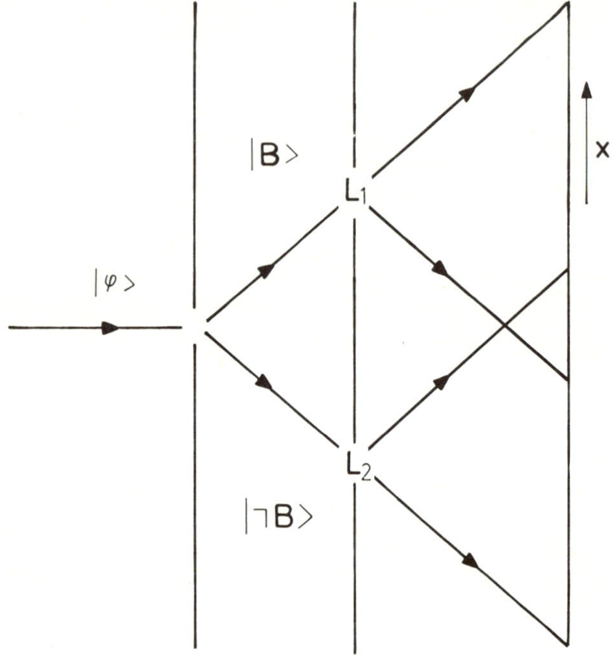

Fig. 12. Schematic representation of the two-slit-experiment.

from which the combined probabilities follow:

$$w_\varphi(A, B) = \langle\varphi|P_B P_A P_B|\varphi\rangle \quad = w_\varphi(B)w_B(A)$$

$$w_\varphi(A, \neg B) = \langle\varphi|P_{\neg B} P_A P_{\neg B}|\varphi\rangle = w_\varphi(\neg B)w_{\neg B}(A)$$

Since in this case the probability is concretely given by $w_\varphi(A)$, one can explicitly calculate $w_\varphi(A)$ with the aid of

$$|\varphi\rangle = \int dx|x\rangle\langle x|\varphi\rangle$$

One thus obtains the probability

$$w_\varphi(A) = |\langle x|\varphi\rangle|^2 = \langle\varphi|P_A|\varphi\rangle$$

and in the general case of incommensurable observables

$$\text{(II)} \qquad w_\varphi(A) = w_\varphi(A, B) + w_\varphi(A, \neg B) + w_\varphi^{int}(A, B)$$

where the interference term is given by

$$w_\varphi^{int}(A, B) = \langle\varphi|P_B P_A P_{\neg B} + P_{\neg B} P_A P_B|\varphi\rangle$$

On the basis of the relation $P_{A \wedge B} \leqslant P_B P_A P_B$, which is valid for arbitrary projection operators, it follows that

(III) $\qquad w_\varphi(A \wedge B) \leqslant w_\varphi(A, B)$

and in conjunction with (II)

(II') $\qquad w_\varphi(A) \geqslant w_\varphi(A \wedge B) + w_\varphi(A \wedge \neg B) + w_\varphi^{int}(A, B)$

If one now objectifies all properties in the sense that they are attributed objectively to the system considered, then one can determine from the logical laws, for which the corresponding propositions are adequate, the probability for the occurrence of a property or the relations between these probabilities.

If one starts with classical logic and the probability calculation characterized by $W_1 - W_5$, then the relation previously deduced[20]

(I') $\qquad w_\varphi(A) = w_\varphi(A \wedge B) + w_\varphi(A \wedge \neg B)$

is applicable; this, together with (II'), immediately leads to the inequality

$$w_\varphi^{int}(A, B) \leqslant 0$$

which is apparently incorrect for

$$w_\varphi^{int}(A, B) > 0.$$

If one, however, uses quantum logic, which is the only reasonable course of action with the assumption of objectifiability, then, instead of (I'), it merely follows that

(Ĩ) $\qquad w_\varphi(A) \geqslant w_\varphi(A \wedge B) + w_\varphi(A \wedge \neg B),$

as has been shown previously. Equations (Ĩ) and (II') are not, however, in contradiction. Moreover, for $w_\varphi^{int} \geqslant 0$ the inequality (Ĩ) follows from (II') and for $w_\varphi^{int} \leqslant 0$ the inequality (II') follows from (Ĩ).

These considerations demonstrate clearly that the objectification of incommensurable properties is not possible if one uses classical logic and the associated theory of probability for propositions on the existence of these properties. Equations (I') and (II') lead to a contradiction. On the other hand, objectification can be carried out if one uses quantum logic instead of logic, and the quantum-logical instead of the classical theory of probability. The quantum-mechanical inequality (II$'$) is compatible with the relation ($\tilde{\text{I}}$) derived from the quantum-logical probability theory. A contradiction then no longer appears.

6. SUMMARY

The considerations of this chapter have demonstrated that effective logic is a theory which is *a priori* valid for all propositions of unrestricted availability. To this type of proposition belong all propositions of classical physics and all the propositions that assert the occurrence of objective properties in a quantum-mechanical system. In addition, because of their special structure, the *tertium non datur*, and hence the complete classical propositional logic, is applicable to quantum-mechanical and classical-physical propositions.

Essential modifications occur, however, if one is concerned with propositions that assert the occurrence of arbitrary measurable, and hence also incommensurable, properties of quantum-mechanical systems. Under these circumstances two cases can ensue: either one makes use of the knowledge of the quantum theory in the laws of logic, and then some statements of logic lose their applicability, but continue to be true; or one explicitly excludes knowledge of quantum theory in the application of the laws of logic, and hence considers fictional objects and their properties. Then some of the laws of logic became false. The reason for this is that the quantum-mechanical propositions are no longer unrestrictedly available, because of the physical conditions under which alone they can be proved, but rather have only restricted availability. The consequence of this is that essential restrictions become apparent already in the effective logic. In particular, the statement $A \rightarrow (B \rightarrow A)$ becomes invalid. The statements of effective logic, which are valid for propositions of restricted validity, we then designated as effective quantum logic. Because of the definiteness of truth values of the propositions used additional statements are possible here. Thus, in particular the *tertium non datur* is also valid in the complete quantum logic. Table VI.1 summarizes the results.

The problem of the validity of logic in quantum physics is closely connected with the questions which emerged in the discussion of the concept of matter and the laws of causality. The problem that is common to these questions is that of the objectifiability of arbitrary properties of a quantum-mechanical system.

Already in the discussion on the concept of matter we demonstrated with the two-slit experiment that the referring of all measurable properties to a system, hence their objectification, leads to logical and probability-theoretical contradictions. The law of causality, as well, loses its validity

TABLE VI.1

Proposition about	Additional knowledge	Logic theory	Validity
classical and quantum-mechanical objects	–	classical logic	w, a
fictional objects	quantum theory	classical logic	$w, \neg a$
	–	classical logic	$\neg w, a$
	–	quantum logic	w, a

Comments to Table VI.1
Column 1 refers to the object dealt with by the propositions,
    2 gives the additional information available for use with the laws of logic,
    3 lists the various theories, and
    4 assesses their validity.
       where: w indicates true, a applicable, $\neg w$ not-true, and $\neg a$ not-applicable.

when applied to incommensurable properties which one interprets as objective properties of a quantum-mechanical system. Finally, our investigation of logic showed that for propositions that assert the occurrence of incommensurable properties of a system, several laws of logic become incorrect. A particular example of such a statement is the logical statement already discussed in connection with the concept of substance. Further, the restrictions of probability theory which emerged in the discussion of the two-slit experiment were revealed to be a consequence of the constraints within logic.

An obvious conclusion from these results is that an objectification of incommensurable properties is not possible. For, as we have seen, objectification leads to logical and probability-theoretical contradictions and to the loss of the law of causality.

But on the other hand, our investigation of logic has demonstrated that by the objectification only a few laws of logic, and correspondingly only a few laws of the theory of probability, are rendered invalid. The law of causality, however, is lost in its entirety. The remaining theories, the quantum logic and the probability theory of incommensurable propositions, are to be regarded as entirely useful logical and mathematical theories which permit a reasonable discourse about the fictional objects of quantum physics.

We can therefore carry out the objectification of all properties, i.e. we can regard quantum-mechanical systems as objects in the classical sense, if we are prepared to accept the mentioned restrictions on logic, the theory of probability and the law of causality.

NOTES

[1] Regarding the question of what is to be more precisely understood by the completeness of a syllogism, refer to G. Patzig, *Die Aristotelische Syllogistik*, Göttingen 1959.

[2] Strictly speaking effective logic is dealt with here. In the following section we shall consider the distinction between effective and classical logic.

[3] The reason for the designations 'restricted' and 'unrestricted availability' will be given in following sections.

[4] In analogy to the concept of effective logic we shall later use the more precise designation 'effective quantum logic'. These concepts will be clarified in the following sections.

[5] In the literature for the operation '→' various denotations have been used. E.g. Lorenzen (ref. [6]) calls this operation 'subjunction', whereas Jauch and Piron (C.f. ref. [11]) use the term 'conditional'.

[6] See, for example, P. Lorenzen, *Metamathematik*, Mannheim 1962.

[7] To counter the obvious objection that the three propositions refer to different times (see, for example, E. Scheibe, *Die kontingenten Aussagen der Physik*, Frankfurt 1964; H. Lenk, *Phil. Nat.* 11 (1969), 413) one should note the arguments advanced above.

[8] Cf. E. W. Stachow, Dissertation, The University of Cologne, 1975; and *Journal of Philosophical Logic*, 1976.

[9] Cf. E. W. Stachow, Diplomarbeit, The University of Cologne, 1973.

[10] P. Mittelstaedt and E. W. Stachow, *Foundations of Physics* 4 (1974), 355.

[11] For the existence and uniqueness of a material implication in the complete quantum logic, see: P. Mittelstaedt, *Z. Naturforsch.* 25a (1970), 1773 and 27a (1972), 1358.

[12] F. Kamber, *Math. Ann.* 158 (1965), 158.

[13] A systematic presentation of the theory of closed linear manifolds of Hilbert space, with particular reference to the logical problem here investigated, was first given by G. Birkhoff and J. von Neumann, *Ann. of Math.* 37 (1936), 823.

[14] J. M. Jauch and C. Piron, *Helv. Phys. Acta* 36 (1963), 827; C. Piron, *Helv. Phys. Acta* 37 (1964), 439; F. Kamber, *Math. Annalen* 158 (1965), 158.

[15] P. Mittelstaedt, *Z. Naturf.* 25a (1970), 1773 (Theorem IV).

[16] For the proof see: G. Ludwig, *Die Grundlagen der Quantenmechanik*, Berlin 1954, pp. 59.

[17] See, for example, G. Ludwig, *Die Grundlagen der Quantenmechanik*, Berlin 1954, pp. 68.

[18] See, for example, J. M. Jauch, *Foundations of Quantum Mechanics*, Addison-Wesley, Reading, Mass. (1968), p. 94.

[19] Because of $w_3^{(q)}$ it follows from the implication $A \leqslant C$ that even for arbitrary incommensurable propositions:

$$w_\varphi(A) + w_\varphi(\neg C) = w_\varphi(A \vee \neg C)$$

which, because of

$$w_\varphi(\neg C) = 1 - w_\varphi(C)$$

leads to

$$w_\varphi(A) \leqslant w_\varphi(C)$$

[20] Formula (I) follows immediately from (I′) together with (III).

BIBLIOGRAPHY

More detailed and specific investigations of the problems discussed in this book may be found in the following works. The bibliography is not claimed to be exhaustive; it should, however, be adequate to serve as a guide to further study. Many of the cited textbooks and papers contain collections of references in their respective fields.

THEORY OF RELATIVITY (Chapters I and II)

Bergmann, P. G., 'Special Theory of Relativity', in *Handbuch der Physik* (ed. by S. Flügge), IV, Springer-Verlag, Berlin, 1962.

Born, M., *Einstein's Theory of Relativity*, Dover Publications, New York, 1962.

Cassirer, E., *Einstein's Theory of Relativity considered from the Epistemological Standpoint*, Open Court, Chicago, 1923.

Einstein, A., *The Meaning of Relativity*, Princeton University Press, Princeton, 1953.

Fock, V., *The Theory of Space, Time, and Gravitation*, Pergamon Press, New York, 1959.

Jammer, M., *Concepts of Space*, Harvard University Press, Cambridge, Mass., 1960.

Landau, L. D. and Lifschitz, E. M., *Classical Field Theory*, Addison-Wesley, Reading, Mass., 1951.

Møller, C., *The Theory of Relativity*, Clarendon Press, Oxford, 1972.

Pauli, W., *Theory of Relativity*, Pergamon Press, New York, 1958.

Reichenbach, H., *The Philosophy of Space and Time*, Dover Publications, 1962.

Reichenbach, H., *Axiomatik der relativistischen Raumzeit Lehre*, Vieweg Verlag, Braunschweig, 1924/65. [English transl. University of California Press, 1970].

Scholz, H., 'Das Vermächtnis der Kantischen Lehre vom Raum und von der Zeit', *Kant-Studien* **XXIX** (1924).

Scholz, H., 'Einführung in die Kantische Philosophie' in *Mathesis Universalis*, Benno Schwabe Verlag, Basel, 1961.

Terletskii, Ya. P., *Paradoxes in the Theory of Relativity*, Plenum Press, New York, 1968.

Weyl, H., *Space-Time-Matter*, Methuen and Co., London, 1922.

Weyl, H., *Mathematische Analyse des Raumproblems*, Berlin 1923.

Weyl, H., *The Philosophy of Mathematics and Natural Science*, Princeton University Press, Princeton, 1949.

QUANTUM THEORY (Chapters III, IV and V)

Belinfante, F. J., *A Survey of Hidden Variable Theories*, Pergamon Press, Oxford, 1973.

Bohm, D., 'Hidden Variables in the Quantum Theory' in D. R. Bates (ed.), *Quantum Theory*, vol. III, Academic Press, New York and London, 1962.

Cassirer, E., *Determinism and Indeterminism in Modern Physics*, Yale University Press, New Haven, 1956.

d'Espagnat, B., *Conceptions de la Physique Contemporaine*, Hermann et Cie, Paris, 1965.

Heisenberg, W., *The Physical Principles of the Quantum Theory*, University of Chicago Press, Chicago, 1930.

Heisenberg, W., *Physics and Philosophy*, Harper and Row, New York, 1966.

Hermann, G., 'Die naturphilosophischen Grundlagen der Quantenmechanik', *Abhandl. der Fries'schen Schule*, Neue Folge 6, Heft 2, Verlag Öffentliches Leben, Berlin, 1935.

Jammer, M., *The Conceptual Development of Quantum Mechanics*, McGraw-Hill, New York, 1961.

Jammer, M., *The Philosophy of Quantum Mechanics*, John Wiley, New York, 1974.

Jauch, J. M., *Foundations of Quantum Mechanics*, Addison-Wesley, Reading, Mass., 1968.

Ludwig, G., *Die Grundlagen der Quantenmechanik*, Springer-Verlag, Berlin, 1954.

v. Neumann, J., *Mathematical Foundations of Quantum Mechanics*, Princeton University Press, Princeton, 1955.

Pauli, W., 'Die allgemeinen Prinzipien der Wellenmechanik', in *Handbuch der Physik*, (ed. by S. Flügge), Springer-Verlag, Berlin, 1958.

Scheibe, E., 'Bibliographie zu Grundlagenfragen der Quantenmechanik', *Phil. Natural.* **X** (1967).

Scheibe, E., *The Logical Analysis of Quantum Mechanics*, Pergamon Press, Oxford, 1973.

Süßmann, G., 'Über den Messvorgang', *Abh. der Bayer. Akad. der Wissensch.* **88** (1958).

v. Weizsäcker, *Zum Weltbild der Physik*, Hirzel-Verlag, Stuttgart, 1960.

QUANTUM LOGIC (Chapter VI)

Birkhoff, G., *Lattice Theory* [American Mathematical Society, *Ann. Math. Soc. Coll. Publ.* **XXV** Rev. Ed.], 1961.

Birkhoff, G. and v. Neumann, J., 'The Logic of Quantum Mechanics', *Ann. of Math.* **37** (1936), 823.

Curry, H. B., *Foundations of Mathematical Logic*, McGraw-Hill, New York, 1963.

Hilbert, D. and Ackermann, W., *Grundzüge der Theoretischen Logik*, Springer-Verlag, Berlin, 1949.

Jauch, J. M., *Foundations of Quantum Mechanics*, Chapter V, Addison-Wesley, Reading, Mass., 1968.

Kamlah, W. and Lorenzen, P., *Logische Propädeutik*, Bibliographisches Institut, Mannheim, 1967.

Kleene, S. C., *Introduction to Metamathematics*, North-Holland, Amsterdam, 1964.

Lorenzen, P., *Einführung in die operative Logik und Mathematik*, Springer-Verlag, Berlin, 1955.

Lorenzen, P., *Formal Logic*, D. Reidel Publishing Co., Dordrecht, 1965.

Lorenzen, P., *Metamathematik*, Bibliographisches Institut, Mannheim, 1962.

Lorenz, K., 'Dialogspiele als semantische Grundlage von Logikkalkülen', *Archiv für math. Logik und Grundlagenf.* **11** (1968).

Mittelstaedt, P., 'Untersuchungen zur Quantenlogik', *Sitzungsber. der Bayer. Akad. der Wissenschf.*, München 1959.

Mittelstaedt, P., 'Quantenlogik', *Fortschr. der Physik* **9** (1961), 106.

Reichenbach, H., *Philosophical Foundations of Quantum Mechanics*, University of California Press, Los Angeles, 1948.

Scheibe, E., *Die kontingenten Aussagen der Physik*, Athenäum Verlag, Frankfurt 1964.

von Weizsäcker, C. F., 'Komplementarität und Logik', *Naturwissensch.* **42** (1955), 521, 545.